Astronomers' Universe

For further volumes:
http://www.springer.com/series/6960

Antony Cooke

Astronomy and the Climate Crisis

 Springer

Antony Cooke
Camino Capistrano
Capistrano Beach
92624 California, USA

ISSN 1614-659X
ISBN 978-1-4614-4607-1   ISBN 978-1-4614-4608-8 (eBook)
DOI 10.1007/978-1-4614-4608-8
Springer New York Heidelberg Dordrecht London

Library of Congress Control Number: 2012941866

Printed on acid-free paper

Springer is part of Springer Science+Business Media (www.springer.com)

Dedicated to my nephew, Stuart, and niece, Georgina, From their loving Uncle Tony, with congratulations to both on attaining their Ph.D.'s in 2011.

Foreword

In February1632 the mathematician and astronomer Galileo Galilei published his *Dialogue Concerning the Two Chief World Systems*. Written at a time when Copernicus' Sun-centered theory of the planets was still relatively new—and highly controversial—it compared that model with the old Earth-centered Ptolemaic picture by means of an imagined conversation between philosophers. The protagonists were Salviati (an erudite Copernican) and Simplicio (a traditionalist, whose name says it all), together with an enthusiastic bystander called Sagredo.

Through several hundred pages of fictional debate, these gentlemen covered a broad swathe of ideas, ranging from the profound to the absurd. While Galileo's undeclared intention was that his readers would make up their own minds whether or not Earth is in motion, it had to be done in a manner that would not arouse the ire of the Holy Roman Church. That all-powerful body was still uncompromisingly wedded to the idea of a stationary Earth, and dissent was dangerous—as had been discovered by the hapless Giordano Bruno some three decades earlier, when he had been burned at the stake.

Although Galileo succeeded in gaining the imprimatur of the Church for his book, he found this to be no guarantee against shooting himself in the foot. Rather incautiously, he had not only aligned Salviati with his own private view but had also associated the clueless Simplicio with the Pope, Urban VIII. This did not go down well in Rome, and Galileo soon found himself facing the Inquisition. The rest is history. On June 22, 1633, he was convicted of 'vehement suspicion of heresy' and, lucky to escape with his life, was sentenced to lifelong house arrest.

While the same fate seems unlikely to overtake the author of *Astronomy and the Climate Crisis*, the parallels with Galileo's *Dialogue* are clear. We live in an era of big questions about our planet. We know its climate is changing. But is that a consequence of human activity? Or is it the result of natural processes that are unrelated to our presence on Earth? It is already clear that the answer to both questions is 'yes,' but the extent to which each contributes to global warming has remained a contentious issue. Despite all attempts to ringfence the science, the arguments have become highly politicized, with powerful vested interests seeking to sway popular opinion.

Into this morass has stepped Antony Cooke with the remarkable volume you now hold in your hands. And he has adopted a strikingly Galilean approach in his investigation of the issues. Wisely, perhaps, he has avoided the 'three-dudes-and-an-argument' format (which is pretty unfashionable these days), but he has followed Galileo in presenting a broad range of ideas and opinions. By introducing an astronomical dimension into the discussion, Cooke has set the climate debate against the widest possible backdrop—itself a major contribution to the literature. And he has not been afraid to tackle the political issues head-on. But, as he declares in the book, it is not his purpose to render a verdict.

There is one other important way in which this book parallels Galileo's *Dialogue*—and that is that they both present snapshots of a rapidly evolving situation. Just as Kepler's laws of planetary motion quickly substantiated the Copernican model of the Solar System, it will soon become abundantly clear which of the many scientific views aired in *Astronomy and the Climate Crisis* come closest to the underlying reality. In that regard, the book represents a valuable record not only of contemporary climate science but of society's treatment of it. It is sure to find a place on the bookshelves of anyone who cares deeply about the environment.

NSW, Australia   Fred Watson

Preface

The science is in, the debate is over!

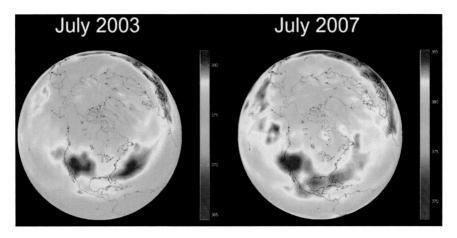

Rising global carbon dioxide levels (note slightly adjusted color gradient from left to right in order to show detail better) (Image courtesy of NASA/JPL)

If only it were that simple; such a grandiose and dramatic pronouncement about any issue would normally mean there is nothing left to discuss. However, with the highly charged nature of the subject of climate change, we will never know if this is so if we do not examine the subject in its entirety, from all sides and perspectives, and free from bias or predetermined conclusions.

There are many books about climate change. However, it is hard to find a source where concise arguments *from all sides* are presented under the canopy of one umbrella, and framed in the context of the actual research behind the findings, much less those that introduce the astronomical component. It is also easy to fall into the trap of mixing science and politics, when, in fact, the two have little in common and objectivity is easily lost. Bearing in mind the late Senator Daniel Patrick Moynahan's perceptive and witty remark that "One is entitled to one's opinion, but not to one's own facts!" let us try to air as many facts and well-founded suppositions as possible. However, we must include at least a brief description of the various agendas relative to the discussion if we are to have a balanced perspective of the broader topic.

> A finding by the Environmental Protection Agency to declare carbon dioxide and other greenhouse gases a danger to public health and welfare – and therefore subject to the Clean Air Act – "will officially end the era of denial on global warming."
>
> –Rep. Ed Markey, D-Mass.,
> U.S. Energy & Commerce subcommittee

The author seeks neither to take a position nor evaluate the science itself, but merely to present an airing of largely unheralded, usually highly legitimate research within the context of the larger umbrella of climate science today. Every effort has been made to present information without a specific agenda or stated conclusions, and especially without the intent of manipulating or swaying the reader to any particular viewpoint. Thus, we will allow the evidence and findings to speak for themselves.

It is quite possible that much of what may be seen as radical at first glance will turn out to be quite compatible with what is already widely accepted. This could mean that perhaps no one will have to reject *all* or even *any* previously held positions. Although the basic theories of man-made climate change have wide international support, there is still much disagreement. So despite oft-repeated claims that the issue is completely settled within the scientific community, to the contrary, this is not yet the case. However, we should be extremely wary of jumping on the bandwagon of those who might perhaps be considered "kooks."

> The underlying cause of these climactic shifts is ultimately not well-understood and is a matter of vigorous debate.
>
> –U.S. Senator Mike Crapo (R-ID),
> Senate Committee on Environment and Public Works, October 2009

With the advent of climate change, aka global warming, taking center stage in the late twentieth century, a worldwide organization called the Intergovernmental Panel on Climate Change (better known as the IPCC) was established by the larger organization of the United Nations. It was structured to review and assess ongoing independent research about the prospects for future impacts on world climate, something that had increasingly come under scrutiny with warmer annual temperatures being recorded. To date, the IPCC has generated four reports: those of 1990, 1995, 2001, and 2007. Each has a designated acronym, i.e., FAR (First Assessment Report), SAR (the Second), TAR (Third), and AR4 (Assessment Report #4). The panel, consisting of members of the international scientific community from 130 countries, has made recommendations with each report about recently changing climate conditions.

The IPCC is comprised of 2,500 expert reviewers, 800 contributing editors, and more than 450 lead authors. This must surely be the largest single organization of its kind in the world. It should be noted, however, that not all of these individuals write everything in the reports, collectively. They each have their own areas of expertise and contribute to the reports accordingly. Climate science, being multi-dimensional, requires input from many of these areas. Therefore, it is unlikely that most panelists are able to be involved in many, if indeed any, of the other areas. This has led to perhaps one of the more serious criticisms that the reports do not speak with one voice. Certainly, no one voice has existed throughout the process.

The panel has not escaped other challenges and resistance to its findings. Critics charge that it has approached the task with a presumption of human-induced recent climate changes. Certainly the IPCC's own mission descriptions do seem to imply this. Critics also challenge its makeup, being merely a small fraction of the world's scientists and experts. We must assume, however, that each country has put forward some of its best and brightest, and there is obviously a limit as to how many people could be involved. Additionally, critics maintain that because any errors it has made over the years have tended to favor its position, this has only added to questions of the validity of its findings.

> Any developed and geographically well-situated nation that is not willing to take action on climate change will have to be fully prepared to assist environmental refugees displaced.
>
> –Australian Greens Senator Sarah Hanson-Young

A careful review of the documents prepared and presented by the IPCC over the years is highly instructive. These reports set the stage and describe in detail many facets of climate change science. Some readers may be surprised to read a concession that many components are still sufficiently uncertain to state definitively. However, the panel has expressed increasing levels of "confidence" that much of what we are experiencing is human-induced (anthropogenic), and that it is exceedingly likely that most of it is not natural.

Interestingly, in the most recent 2007 report (AR4), the concluding remarks may seem surprisingly short of the kind of definitive statements about anthropogenic factors that we have been led to believe were at its core. In stark contrast, however, is the *Summary for Policy Makers* that accompanied the report. This states far more clear-cut conclusions than are found in the main report. Whether this was the intent or the result of a less objective editor is not clear. One does not leave the summary without a strong sense that the most likely or worst-case scenario has been stated as a virtual certainty. Critics have been vocal about the summary for this very reason, especially since the document has been at the root of many decisions and actions taken by governmental agencies.

> The government has been captured by an extreme Green agenda...incomplete and biased science.
>
> –Australian Liberal Party senator
> Cory Bernardi

In light of a prevailing view (a "consensus") that man-made contributions have caused a large portion of the warming since 1980, it does seem reasonable to presume that a large percentage of scientists do indeed accept much, if not all, within the IPCC's findings. We have all heard those official-sounding ratios of scientists in support of the IPCC position to be in the upper 90s percentile. Indeed, many major scientific institutions of international standing have affirmed their support of the IPCC assessments. Apparently, to date, none has gone on record opposing it, although this is not quite the same thing as claiming all institutions actually have embraced it, let alone having spoken for everyone working under their collective umbrella. Without precise headcounts of individual scientists, it surely would be hard to gauge what any of them think, individually, as there has been no census taken. However, as we proceed, statements given in absolutes by advocates of any position should be carefully appraised.

The legend surrounding the early English King Canute comes to mind. Enthroned on the seashore, the king of legend (and of history) attempted to command back the tides. Needless to say, he ended up with wet feet. However, just for the record, apparently he was wiser than his legend has fared. It seems his intent was to demonstrate that no person, no king, had such power. Thus, Canute demonstrated to an unwitting peasant populace that it is impossible to impose anything on the natural order of things.

About the Author

It has been said that music and astronomy go hand in hand. Antony Cooke's passion for both fields was clear very early in his life, but music ultimately would claim his career. A cellist of international renown, Cooke has been one of the leading players in the Hollywood recording industry for many years, having been associate professor of cello at Northwestern University in Chicago until 1984. He is a US citizen but was born in Australia and educated in London, receiving numerous prizes and awards, including the Gold Medal at the London Music Festival.

The dual nature of Cooke's interests continued, astronomy remaining the counterbalance in his life. Always looking for ways to improve his experience at the eyepiece, he has constructed many telescopes over the years, with increasing sizes being the hallmark of his (often-quirky) designs. He has published four previous books for Springer, *Visual Astronomy in the Suburbs* (2003), *Visual Astronomy Under Dark Skies* (2005), *Make Time for the Stars* (2009), and, recently, *Dark Nebulae, Dark Lanes and Dust Belts* (2012).

Contents

1. The Astronomical Connection

Anyone who doubts that there is any point in discussing any further the matter of climate change should take a look at NASA's own website [1]. The section entitled *Uncertainties: Unresolved Questions about Earth's Climate* disavows any sense that we do indeed know everything. While described almost casually as "just a few other important uncertainties about climate change," it is a concise description of many hugely significant factors often cited by some to be understood with certainty! It is not as if NASA has minimized these issues, but they do point to a possible connection to astronomical research. And because the space agency is heavily involved in climate research, we should realize that astronomy is a growing part of it.

> Halting global warming requires urgent, unprecedented international cooperation, but the needed actions are feasible and have additional benefits for human health, agriculture and the environment.
>
> James Hansen, 2003
> Director, Goddard Institute of Space Sciences

We should always keep in mind that the focus of most mainstream climate change research revolves around human-induced causes of climate change (AGW, Anthropogenic Global Warming), and the corresponding increased levels of atmospheric carbon dioxide (CO_2). Michael J. Dougherty, in an eloquently stated and wide reaching opinion piece [2], noted that climate science is a multidisciplinary field, which includes ecology, chemistry, geology, glaciology, meteorology, atmospheric science, marine biology, volcanology, computer modeling, as well as many other disciplines. All in all quite a complex recipe, and one that has seen its fair degree of heated dialog exchanges. However, Dougherty, clearly convinced of AGW theories, made great efforts to be

A. Cooke, *Astronomy and the Climate Crisis*,
Astronomers' Universe, DOI 10.1007/978-1-4614-4608-8_1,
© Springer Science+Business Media New York 2012

conciliatory and fair to scientists of every view. In assessing and stating his position, his article is not out of line with similarly fair-minded proponents.

However, maybe Dougherty will allow us one little caveat. Upon reading the various scientific fields that he lists, it is interesting that one discipline has been omitted from the mix – astronomy. We can only presume this is not a deliberate slight. It is probably not an oversight, either, but a genuine reflection of how he sees the makeup of applicable fields in climate research. However, it is striking that the one missing element is seen increasingly by a growing number of researchers as a significant component of climate science. It will remain the focal point of this book.

> All the evidence I see is that the current warming of the climate is just like past warmings. In fact, it's not as much as past warmings yet, and it probably has little to do with carbon dioxide, just like past warmings had little to do with carbon dioxide.
>
> William Happer, 2009
> Princeton University

Some serious scientists pursuing possible astronomical connections to climate change effectively find themselves questioning the IPCC position discussed in the preface of this book. Unfortunately and by default, they remain tied to and somewhat unkindly lumped in with all climate "skeptics." Others have complained that the IPCC has left astronomers and astronomy out of the equation, which does seem fair in light of the panel's scant reference to most studies that theorize astronomical links to climate. Still more scientists believe their more recent studies have demonstrated convincingly that astronomical links to climate are indisputable. Whether the IPCC ultimately deems them to be valid or not, it seems they have barely been referenced at all.

As we proceed, we will examine the underlying fundamental physics and mechanics – known and projected – of climate itself. These are essential to our understanding of the greater picture. But also we will look to the stars. A sizeable volume of serious climate research has implicated forces beyond our local Earthly environment, information that generally has not been presented to the public at large. Additionally, there are numerous peer-reviewed astronomical/climate studies in many of the most important and relevant of scientific journals that encompass often-startling concepts. Even the IPCC acknowledges certain astronomical tenets (albeit limited in scope).

Although not every researcher is looking to the skies for external explanations for recent changes in the climate, certainly every scientist understands how Earth's place in the cosmos indeed does impact the conditions in which we live. The question is whether such external factors have any role in the observed warming in the late twentieth century. It is a good question, even though many have already concluded that such factors do not. However, because the recent warming does not appear to fit normal patterns, it is reasonable that it be investigated for any possible explanation or factor.

Obviously, astronomical research into climate includes the variable output of the Sun, what may be behind it, and what this may or may not imply for Earth's climate. Additionally and more interestingly, there are the Sun's possible indirect effects on climate, other than the direct heating processes we already know about. Further research has been done concerning other possible local factors and possible secondary responses from influences far outside our environment, ultimately affecting the climate (and producing tangible effects on water vapor concentrations, cloud cover, melting icepacks, changes to ocean movements, heat storage, etc.).

Within the Solar System, studies have included the potential influences of cyclic, orbital and precessional effects, not only of the Sun/Earth/Moon system but all of the planets together and separately. This includes possible forces and events we do not yet understand. Further, there are solar magnetic flux variations that may control certain complex mechanisms, and their theorized but yet unproven effects on Earth.

Beyond these, we cannot blindly ignore without investigation even what may seem to be truly farfetched – the effects of cosmic rays from deep in the Milky Way Galaxy, possible consequences of the Solar System's place in the galaxy, or movement through regions of meteorite debris, dust, or dense interstellar matter. Could there be other yet undetermined galactic forces, or the interactions of these with any other factor(s) previously mentioned? Most of these have been subjects of numerous research papers.

> How did Earth, Venus and Mars wind up so radically different from similar origins? How could Mars have once been warm enough to be wet, but be frozen solid now?
>
> These questions revolve around climate and the intersection of climate, atmospheric chemistry and, on Earth, life.
>
> NASA

We can speculate, too, on the possible result of a galactic merger, whereby we might find ourselves in the midst of starburst activity, with the accelerated, super-heated star cycles that are known to result. At least in regard to recent climate warming, such starburst theories can readily be set aside, as the research paper by Brian. A. Keeney et al. aptly demonstrates [3], although it makes for fascinating reading. We will explore whether any of the remainder can be tied to the observations of recent climate change on Earth or if they remain just fanciful ideas in search of legitimacy.

Most of these scenarios presently suffer less from being unproven than to being considered as even remotely possible ties to recent warming trends. Remarkably, the body of astronomical climate research outside the mainstream still seems barely known to the public, if at all. Those searching for connections to the stars include scientists of various descriptions, even scientists of every discipline, who have found their original areas of expertise increasingly tied to the field of climatology. Many have questioned the degree of influence of human-induced factors, resulting in questioning from a number of schools of thought:

- Is AWG (anthropogenic global warming) *all* that is responsible?
- Is AWG even responsible *at all* for what has been observed over the past few decades?
- Or is it a *combination* of factors, perhaps, of both human-induced and natural processes, including even astronomical factors?

Where to Start

Numbered references to all key research papers, other sources, articles, or websites mentioned in the book (according to the sequence presented in each chapter) are included at the end of each chapter. It is strongly recommended that readers avail themselves of these references as much as possible, because the kind of in-depth attention they deserve cannot be included under the cover of just one book. Most can be found online and have been selected from as broad a perspective as possible. Wide as it is, though, it represents a mere fraction of all that is available.

However, for those wishing for a more detailed background on mainstream climate science than can be provided here, an excellent source would be *Climate Change: A Multidisciplinary Approach* by the late William James Burroughs [4]. A leading climatologist and author, Burroughs brings a lifetime of expertise to this well-written and documented text. Its balanced, wide-reaching approach, with fine presentations of scientific principles, always respectful in its analysis of the theories of others, especially concepts unproven at the time of writing (2001) is still largely just as valid today. It is also a far-reaching commentary that includes touching on a few usually ignored astronomical possibilities.

It is worth mentioning that in 2003, since the writing of his book, Burroughs – the extreme moderate – went on record with the opinion that the IPCC had given insufficient credence to the Sun's role in recent climate warming trends. But now we start by looking at Earth's place over the course of far distant history in order to have a proper perspective of the basic facts that frame what has been experienced over the last 100 years.

Our Place in Time

The time in which we live is termed the Holocene Period, which began about 11,700 years ago. (*Holocene*, from the Greek language, means "most recent.") It marks a period of warming associated with the present interglacial period, extending through today. The beginning of the period marked the point at which glaciers began their retreat, releasing huge burdens of weight from the landmasses below.

Now look at the climate statistics shown in Fig. 1.2, drawn from multiple sources, most of them "proxy" sources – indirect methods of measurement, a system that is regularly used to remarkably good effect in reconstructing the past. Circumstances known to accompany certain stimuli provide conditional reliability of such projections, in this case, tree rings, ice cores, sediments, etc.

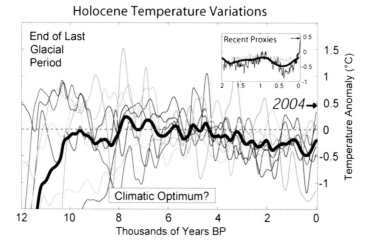

FIG. 1.1 Earth temperatures over the last 12,000 years (Graph by Robert A. Rohde, courtesy global warming art project, prepared from multiple sources)

Their value is because the historical record extends over a period of time far greater than any known manmade records, although for many reasons, there is seldom uniformity of results from the different types of proxy record. This may be seen from Fig. 1.1, but clearly an overall picture can be assessed with a fair degree of confidence. The thick black line represents an average of all inputs, since none, on its own, can be considered completely dependable. Would you think on first glance it shows a scenario that implies any kind of immediate problem?

However, note where the "2004" arrow is positioned relative to the curve average. Suddenly the graph tells a different story. Carelessly, one could thus easily draw an incorrect conclusion (indeed, some less than honest individuals did so deliberately with this very graph). Because it does not show incremental variations of less than 300 years, dramatic temperature increases on short time scales do not register. Therefore it is easy to see how graphs such as these have been manipulated to show different things to misrepresent what may really be happening.

Overall, during the course of the larger 12,000-year period, evidence indicates temperatures today that are decidedly cooler or warmer than they were many thousands of years ago, but significantly over the last 4,000 years they have averaged below the

0°C anomaly threshold. Be sure to look carefully at the amounts of temperature increase, too (temperature anomaly readings to the right) in assessing the present situation and the various records. Reasons for these radical long-term changes in our climate have been largely accepted to include far-reaching astronomical factors. However, the time scale is what sets them apart from the recent phenomena.

Visible Signs of Warming

We have all heard and read about the potential consequences of ongoing climate change – ever higher temperatures, glacial retreat, sea level rise, inundation of coastline cities, water shortages, increased storm intensities and frequencies, crop loss, food shortages and the spread of diseases, to name just some. Actually, most of these are far from unprecedented, having taken place repeatedly before (with one or two notable exceptions – such as the inundation of cities – of course) over a time frame of millions of years.

Meanwhile, the dialog is intensifying. As will be seen to be the case, there are always some views contrary to those seemingly securely established, and others that move the debate to new places. At the present time, for example, there are some scientists who believe we are entering a period of prolonged *cooling!*

We should not confuse weather with climate; one refers to the near short-term, and the other to long-term *prevalent* conditions. Forecasting the weather from day-to-day, or week-to-week, is not subject to the same parameters, even though the two larger fields are related. Climate change, though, is a far longer-term affair. Determining precise measurements of the extent of cooling in the upper atmosphere (considered integral and opposite to that of climate warming in the lower atmosphere) might be a good indicator, although there have been difficulties in doing this in the past. Methods have improved over the past decade, though. Before the advent of a new laser technology (the LIDAR system), these measurements were limited to equipment on board scientific balloons at an altitude of not more than 20 miles [5]. Here is one of the indisputable benefits of NASA funding and research, and of course, a direct benefit of space technology in looking outward.

FIG. **1.2** Glacial retreat; the Helheim Glacier in Greenland (Images courtesy NASA/GSFC/METI/ERSDAC/JAROS, and the U.S./Japan ASTER science team; created by Jesse Allen, using data from NASA's Terra Satellite)

First, though, we should look at some of those indicators that are clearest – these are iconic examples, indeed. We need to look for any reason to implicate factors beyond Earth's immediate environment that could impact any of them. Compounding matters, reversals of any observed climate trend or indicator may take a while to register with any significance; Earth has quite a powerful natural buffer against sudden and radical change. Astronomical studies have been able to compare Earth with other planets in this regard, although the contrasts are stark indeed! Remarkably, even if Earth did enter a prolonged period of cooling, natural repair to some of these obvious indicators would not necessarily show up for decades.

Perhaps the best-known symbol of a warming climate is glacial retreat. Notwithstanding the notorious 2011 investigation of a U.S. Interior Department scientist for potentially manipulating polar bear images in relation to melting glaciers [6], the image (Fig. 1.2), dramatically illustrates what is occurring throughout the world. With the exception of a few glaciers outside of Scandinavia, the rate of melting seems to be intensifying. Most striking are the rapid glacial retreats in Greenland and the western regions of South America.

Certain researchers have claimed that large soot particles from incomplete combustion may, in fact, be more responsible for the melting of glaciers than higher temperatures. Speculation has emerged that this is due to the decreased reflectivity of formerly clean snow and ice and the heat absorbing properties of dark, non-reflective matter coating it. Regardless, the shrinking of these frozen regions is a reality, and soot is certainly a by-product

YBP Library Services

COOKE, ANTONY.

ASTRONOMY AND THE CLIMATE CRISIS.

Paper   280 P.

NEW YORK: SPRINGER, 2012
SER: ASTRONOMERS' UNIVERSE.

LCCN 2012-941866
  ISBN 1461446074   **Library PO#** GENERAL APPROVAL

		List	44.95	USD
5461 UNIV OF TEXAS/SAN ANTONIO	**Disc**	10.0%		
App. Date 1/16/13 PHY.APR 6108-11	**Net**	40.46	USD	

SUBJ: 1. ASTROMETEOROLOGY. 2. CLIMATIC CHANGES--
ENVIRONMENTAL ASPECTS.

CLASS QB28   DEWEY# 520.   LEVEL GEN-AC

YBP Library Services

COOKE, ANTONY.

ASTRONOMY AND THE CLIMATE CRISIS.

Paper   280 P.

NEW YORK: SPRINGER, 2012
SER: ASTRONOMERS' UNIVERSE.

LCCN 2012-941866
  ISBN 1461446074   **Library PO#** GENERAL APPROVAL

		List	44.95	USD
5461 UNIV OF TEXAS/SAN ANTONIO	**Disc**	10.0%		
App. Date 1/16/13 PHY.APR 6108-11	**Net**	40.46	USD	

SUBJ: 1. ASTROMETEOROLOGY. 2. CLIMATIC CHANGES--
ENVIRONMENTAL ASPECTS.

CLASS QB28   DEWEY# 520.   LEVEL GEN-AC

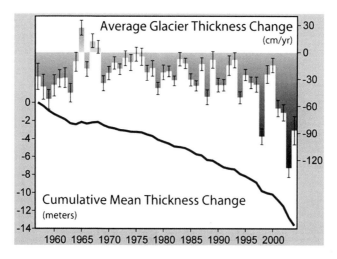

FɪɢG. **1.3** This chart illustrates not only changes in thickness of these massive ice reserves but more importantly the significant increase in the rate of retreat since the early 2000s (Graph by Robert A. Rohde, courtesy global warming art project)

of some fossil fuels. However, this poses a new wrinkle previously not taken into account. We have observed similarities in surface ice and volcanic ash deposits on other Solar System members, and such studies can aid those on Earth.

All may not be what it seems at first glance, however. Glaciers have come and gone almost countless times before between past ice ages. That the present retreat can be directly tied to increasing temperatures is certainly not in general dispute. However, many scientists have pointed out that glacial retreat was well under way long before the recent warming period, commonly attributed to human-induced warming. Their retreat began at a time when the rise in carbon dioxide (CO_2) levels was virtually insignificant, (circa 1850), remaining so until well into the twentieth century. These researchers have instead attributed such phenomena more to our gradual emergence from a cold period in the recent Holocene, better known as the "Little Ice Age" (see Chap. 3), than to CO_2, a factor that has astronomical ties.

Records seem to indicate that the average rate of glacial retreat into the early 2000s was virtually the same as it had been since the mid-1800s – even reversing the trend slightly during mid-century (Fig. 1.3). However, new studies have indicated an accelerated pace has been taking place since 2003. It has been claimed this will

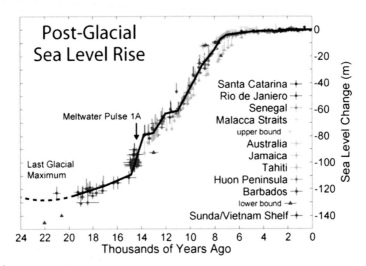

FIG. 1.4 Sea level increases, as reflected in numerous proxy records (Graph by Robert A. Rohde courtesy global warming art project)

result in an increase to sea levels by 1.3 mm yearly [7]. Again, human activity has been held to blame, but the statistics and projections remain confusing and controversial, nevertheless. Wrong or right, some scientists are looking to the skies for answers.

The consequences of the melting of glaciers reaches far beyond those that are obvious. Easy to see is the potential for sea level rise, which indeed has been measured. Obviously, an increase in sea level of just a meter or two would be a serious threat to coastal cities and low-lying regions, so we ought not to dismiss the potential dangers of such prospects. Entire countries, such as Holland, could be threatened by sea level encroachments taking place beyond their ability to defend against them.

The graph (Fig. 1.4) shows the steep increase in sea level that has occurred over the past 24,000 years. It reaches back well before the present Holocene Period, but we can trace where the Holocene began on the graph (at about 11,700 years ago). Up to this time, considerable land compression on the continents had been caused by the massive weight of glacial ice. As it melted, the rapid rise of those landmasses over the first few thousand years is clear on the graph, mimicking a rubber block that has been compressed and released. Thus the land was approximately 120 m lower, and coastlines were often surprisingly well inland from today's coastal regions, explaining why oceanic fossils are frequently found deep inland.

On the scale of the graph, which is admittedly coarse, any recent or present increases seem slight, and do not look to be more than a nominal trend. Most of the major changes already occurred more than 5,000 years ago. However, one should not be easily deceived by appearances, since the differences in present and possible future sea level rises – and disaster for low-lying areas – are only a meter or two. Only time will tell what the final outcome turns out to be for sea level increases in the twenty-first century. If we are prepared at this stage to accept the recent projections of those who have been wrong on so many occasions before (!), we should bear in mind that estimates have been repeatedly revised *downwards* over the course of successive previous IPCC assessments.

Despite ongoing disputes about how much or how little sea levels are actually rising, it seems, regardless, that almost no serious scientist would dispute that it is occurring. We will explore some theories that examine possible links to factors beyond our immediate environment.

Another graphic from NASA reveals a projection for New York City for 2050 and beyond (Fig. 1.5). NASA claims such increases could amount to as much as 40–65 cm by 2100, but critics continue to claim this amount has been greatly exaggerated (see Chaps. 3 and 8). This demonstrates the difficulties of using projections of "chaotic" factors in climate models and extends to knowing which factors even to include (external or internal) in them.

On another front, deforestation has been occurring in various parts of the world for a long time, but never has the trend been taking place to the extent that it is now. This has been one of the issues at the forefront of climate change discussions, not only because of the reduced capacity of the land masses of Earth to absorb carbon dioxide and heat (see Chap. 2) but also the huge accompanying release of carbon compounds into the atmosphere from the destruction of the forestry itself. Additionally, changes to water vapor concentrations and to Earth's albedo (see also Chap. 2) are profound. Here, at least is one area that all scientists can agree upon completely, especially since it is virtually all caused by human activities. Although astronomical science may not be directly related to human activity, indirectly certain common climate variables on other planets can provide some clues, since we can examine the atmospheric effects of different atmospheric concentrations of CO_2 on them.

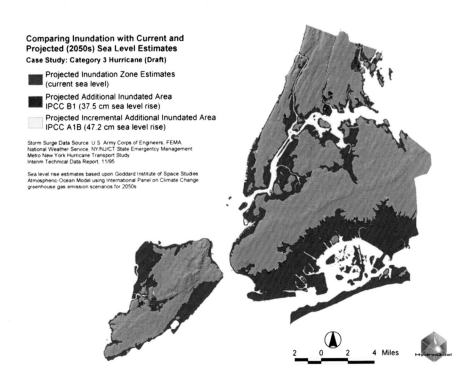

Comparing Inundation with Current and
Projected (2050s) Sea Level Estimates
Case Study: Category 3 Hurricane (Draft)

Projected Inundation Zone Estimates
(current sea level)

Projected Additional Inundated Area
IPCC B1 (37.5 cm sea level rise)

Projected Incremental Additional Inundated Area
IPCC A1B (47.2 cm sea level rise)

Storm Surge Data Source: U.S. Army Corps of Engineers, FEMA,
National Weather Service, NY/NJ/CT State Emergency Management
Metro New York Hurricane Transport Study
Interim Technical Data Report, 11/95

Sea level rise estimates based upon Goddard Institute of Space Studies
Atmospheric-Ocean Model using International Panel on Climate Change
greenhouse gas emission scenarios for 2050s

2 0 2 4 Miles

FIG. 1.5 Projected sea level for New York City, 2050 (Graphic courtesy NASA)

Earth's Lungs

A common perception, frequently used to describe the Amazon rainforest, is that it acts as the "Lungs of our Planet," or "pollution filters." As one of the world's greatest natural resources, the Amazon rainforest continuously recycles carbon dioxide into oxygen, providing about 20% of Earth's oxygen.

Most of this is really not in dispute, but a common misconception is that the rainforest can counteract our excess CO_2, something now being lost through deforestation. And because CO_2 is not a pollutant in the normal sense of the term, we should not regard it as a filter. However, the mature rainforest does store a considerable amount of carbon (in full saturation as a "carbon sink"), and maintains a balance between the CO_2 it takes in and the resulting O_2 it releases. However, we should not regard the shrinking rainforest as a lost panacea to counteract an excess production of CO_2.

Earth, however, has the distinction in the Solar System of having vegetation and continental forestry. This must be factored into the total equation, and related observations of other planets can help us understand the effects of varying albedos, despite their lack of plant life. Although vegetation represents a substantial part of the system that reflects or absorbs energy, it also converts carbon dioxide back into breathable oxygen. Vegetation also recycles water vapor; without this critical component in the atmosphere, regions quickly become arid, since there is no existing air saturation to stop the moisture loss. In contrast, the decreased heat absorption caused by deforestation of tropical regions may increase reflective properties of the regions, thus providing a net cooling effect! At ever-higher latitudes, snow cover will increasingly reduce heat-absorbing properties of forests; thus deforestation in these regions actually has a lesser impact than it does at the tropics. Similarly, desert regions expanding at the expense of forestry might have a surprising net cooling influence on the climate as a whole! We can see the complexities in trying to build appropriate factors into reliable climate models.

Although some regions have been able to claim net gains in forestry during recent times, (especially the United States), there is overall a large total loss worldwide to agriculture and timber harvesting, even to land speculation and unplanned city sprawl. Many less-developed countries exercise no control whatsoever on any of these. The satellite images shown in Figs. 1.6 and 1.7 illustrate the problem well. Studies of other members of the Solar System also serve us well by showing us how landscapes are affected by the lack of vegetation.

We can take this a little further by examining the NASA graphic (Fig. 1.8), which shows the extent of remaining world "tall canopy" forestry (in which trees are typically 30–45 m in height). These regions also contain the most prolific biodiversity – the greatest wealth of living organisms. We can see that there is a large extent of such forestry remaining in South America (although it is not the tallest), but at the current rate of loss it could soon resemble Africa. Since only "tall canopy forests" are shown, wherever the graph appears blank (white) the graphic should not be interpreted as showing a lack of *any* type of vegetation. However, it should give rise to the question of how long it will be before there

FIG. **1.6** Mato Grosso State in Brazil, 1992 and 2006; Brazilian deforestation continues at a rate of 20,000 km^2/year (Image NASA/USGS Landsat Mission; Courtesy NASA/GSFC/METI/ERSDAC/JAROS, and the U.S./Japan ASTER Science Team)

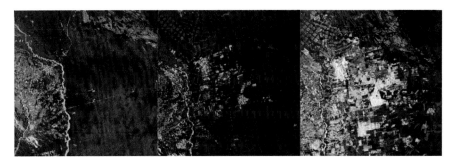

Fig. 1.7 Dry tropical forest region east of Santa Cruz de la Sierra, Bolivia (Image courtesy NASA/USGS; Landsat mission)

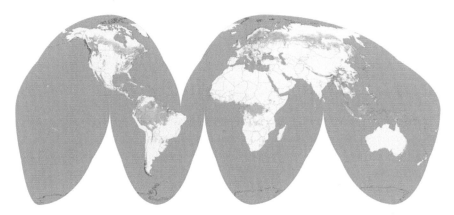

Fig. 1.8 Earth's remaining tall canopy forestry in 2010 (Image courtesy of NASA earth observatory; map by Jesse Allen and Robert Simmon, based on data provided by Michael Lefsky)

is no tall canopy forestry left anywhere at all, and much of Earth's dry land could end up looking like Mars.

Much of the decline in forestry came about during the twentieth century. Additionally, one has to appreciate the huge release of CO_2 from burning all that was not harvested from the deforestation itself, in addition to that emanating from the remaining vegetation that has been allowed to decay on site. It has been reliably estimated that the release of human-induced CO_2 into the atmosphere by deforestation and burning now might have reached as much as possibly 30% of the total, versus just 18% in just the recent past. Although this includes using wood as fuel, it does not mitigate the result.

Complicating matters, the release of soot and other products of combustion (in the form of tiny droplets of sulfates, nitrates, ammonium compounds, and even dust into the air as aerosols) can produce a certain amount of cooling through their reflective rather than absorptive properties! Volcanoes have also played a part in ejecting sulfate aerosols into the stratosphere (at altitudes up to 50 km), in contrast to the elevation of most other aerosols reaching only the troposphere, (not more than 20 km up). Release of these aerosols can also result in cooling trends that last for a period of years because at higher (stratospheric) altitudes these particulates stay aloft for far greater periods of time. Obviously volcanic aerosols are an unpredictable factor in climate change because of their relatively sporadic nature. They must have figured far larger in ancient history, when volcanic activity was greater than it is now. However, some recent, early twenty-first century major volcanic events may herald effects not yet tabulated. Again, this ties into certain space and astronomical studies.

A Happy Ending?

The "Gaia" hypothesis (after James Lovelock [8]) may leave what could seem a depressing subject with a note of optimism. This hypothesis is truly the ultimate wild card. Simply stated, it refers to a philosophic/scientific philosophy that proposes that the cosmos, world and the life it shelters are all interrelated. Thus, no matter what occurs, it is always balanced by another reactive mechanism that maintains the habitability of the environment.

To some degree, we can witness such forces almost every day, where nature seems to have a response for almost any calamity, whether it be the oceans' clean-up of oil spills or other types of disaster, manmade, external or otherwise. However, before we become too cheered by such rosy thoughts, in his later work [9], Lovelock concedes that man could quite possibly overstep the boundaries of Earth's habitat – beyond the point that nature can protect against.

The Complexity of the Subject

Because it is not hard to find evidence of climate change, this is not where the argument ought to be centered. Earth has definitely warmed (by approximately ±0.8 °C, depending on which record is used) over the past century. However, questions have emerged about whether the upward curve of warming that was observed in the 1990s will continue, or has even been sustained. Therein lies the challenge, and especially as some scientists try to determine if astronomical factors have played any part in it, or might be expected to do so.

Let us not underestimate the immense difficulties of studying this subject. There are so many factors in play that even predicting the daily weather by today's meteorological computers requires trillions of calculations. Additionally, difficulties remain in identifying and measuring the effects of human/climate/astronomical interactions, understanding the role and influences of long-term climate cycles, as well as their compounded effects, what is responsible for these cycles, separating those seemingly random components usually regarded as "noise" in the signal, as well as those markers that do not relate to any factor.

Because much of what we experience in the universe is "chaotic" in nature, it is impossible to predict total outcomes. This is the opposite of "linear" systems, where clear cause and effect is traceable and identifiable. Although we may have the answers to all of these factors definitively established one day, there can be no denying that the job seems daunting by any standards. However, some exciting new research from the astronomical community is at very least worthy of our attention.

References

1. Unresolved questions about Earth's climate. Global Climate Change, NASA. Online: http://climate.nasa.gov/uncertainties/
2. Michael JD (2009) Can science win over climate change skeptics? ActionBioscience (July) Online: www.ActionBiosciene.org/education/dougherty.html
3. Keeney BA, Danforth CW, Stocke JT, Penton SV, Shull JM (2005) Does the Milky Way produce a starburst wind? In: Proceedings of the International Astronomical Union, vol 1, pp 424–426
4. Burroughs WJ (2001) Climate change: a multidisciplinary approach. Cambridge University Press, Cambridge
5. Kloeppel JE (2001) Upper atmosphere temperatures hold some surprises. Daily University Science News. http://www.unisci.com/stories/20011/0320015.html
6. Siegel K (2011) Putting an Arctic scientist on ice. Huffington Post (11 Aug 2011). Online: http://www.canadaimmigrationblog.com/kassie-siegal-putting-an-arctic/-scientist-on-ice?Iang=sk
7. Alan B, Janet W (2011) NASA warns ice melt speeding up. NASA (9 Mar 2011). Online: http://www.jpl.nasa.gov/news/news.cfm?release =2011-070
8. Lovelock J (1979) Gaia: a new look at life on Earth. Oxford University Press, Oxford/New York
9. Lovelock J (2009) The revenge of Gaia: Earth's climate crisis and the fate of humanity. Basic Books, New York

2. The Physics of a Crisis

Earth and Space

In spite of the differences and controversies about recent warming trends, there is one common denominator that ought not to garner any disagreement from anyone. Despite often heard claims that the Sun is not the cause of warming, in fact, just a little thought soon makes clear that actually it is present in *every* scenario, no matter how we look at things. It is indeed responsible for all but the minutest energy from other distant sources in the galaxy, or from the even remoter cosmic microwave background (Fig. 2.1). Thus, one way or another the Sun is always part of any climate scenario, whether directly, indirectly, or by any other less obvious means. If it were not for its energy, life could not exist. And we need to recognize that climate change can mean cooling instead, depending on the larger principles governing it. Things do not always have to get hotter.

Of the cosmic microwave background, precise measurements of extremely slight radiant energy allows us to peer deep into the farthest observable universe at the true remnants of the Big Bang itself, akin in layman's terms to glowing embers in a long extinguished fire. From this, we can readily determine the fractional energy spectrum remaining from any part; the colors of the illustration here are not actually true to the temperatures themselves, as the maximum shown represents a "tropical" 2.725 K, which is close to absolute zero. So marshmallows would likely take a while to toast…

At these temperatures there is, of course, no visual light as it appears here in this mapped image, but it helps to make the point. Interesting, too, is the remarkable unevenness of the distribution of energy, but that is a story for another book.

A. Cooke, *Astronomy and the Climate Crisis*,
Astronomers' Universe, DOI 10.1007/978-1-4614-4608-8_2,
© Springer Science+Business Media New York 2012

Fɪɢ. **2.1** The cosmic microwave background temperature fluctuations from the 7-year Wilkinson Microwave Anisotropy Probe data seen over the full sky (Image courtesy NASA/WMAP Science team)

Climate Mechanics

It will not be possible to proceed in any meaningful way unless we have some background of the principles underlying the mechanics of mainstream climate change. It is necessary to understand these theories to understand how they align or contrast with the various astronomical scenarios that are the focus of this book, and if presented in isolation would be almost meaningless. This chapter also offers some comparisons of different positions, since confusingly, there are often numerous interpretations of the same components.

The foundations laid, later chapters can focus on matters that some readers might not have been aware exist at all, perhaps having already concluded that there was nothing left to discuss or examine in broader detail. This involves substantial legitimate alternate, or differently oriented research into all aspects of climate change science.

The Greenhouse Effect

A starting point is to understand how electromagnetic energy from the Sun interacts with Earth and its surrounding atmosphere. What is not reflected is absorbed, re-radiated and compounded in a sequence of events known as the greenhouse effect. We should first understand what this actually is, as it is much misunderstood and usually only implies a negative phenomenon.

Inviting a parallel with glass greenhouses, the analogy has always been incorrect. Glass buildings derive their warming effect from the elimination of convection, and thus the literal trapping and heating of air confined to a small volume of space is responsible for the true greenhouse warming effect. Because this air is warmed by heat radiated directly from the ground below, it cannot re-radiate heat to the air all around it on the outside – actually, an entirely different process to that underlying that of the misnomer.

A common misconception is that the greenhouse effect is a recent phenomenon, caused entirely by increasing concentrations of carbon dioxide. In fact, we depend on it for our survival. Its largest constituent, water vapor, has ensured that the temperature of our environment stays within a livable range. The entire disagreement on climate change revolves around whether humans are artificially increasing Earth's greenhouse temperatures to unlivable levels.

What is not directly reflected back into space is absorbed and re-radiated by Planet Earth and its atmosphere. Only about 6% of the total solar energy absorbed is re-radiated directly back into space in the form of infrared wavelengths. The atmosphere absorbs the majority of the re-radiated energy, adding to the basic warming through a process of the subsequent continued re-radiation of infrared energy to molecules of surrounding atmospheric gases. Energy is thus re-radiated back and forth between molecules in an almost endless chain until all is eventually lost to space. However, more solar radiation continues to enter the environment. A balance is achieved when the lag between absorption and total loss keeps Earth at a habitable temperature. Despite this circular process, obviously, if further incoming radiation were to be blocked, all existing energy eventually would be dissipated away from Earth.

Earth's overall temperature is not a given constant and can change as a result of a variety of factors. Some might well be avoidable, and others not. The concern today is that warmth is being absorbed faster than it can be re-radiated to maintain a livable environment, because of additional greenhouse gas contributions. Figure 2.2 illustrates the basic mechanisms of the greenhouse effect, minus the influence of some variables, such as clouds and aerosols.

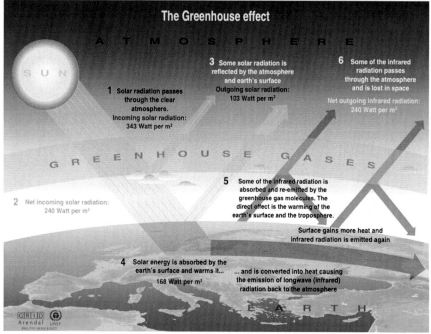

Sources: Okanagan university college in Canada. Department of geography. University of Oxford. school of geography. United States Environmental Protection Agency (EPA). Washington: Climate change 1995. The science of climate change. contribution of working group 1 to the second assessment report of the intergovernmental panel on climate change. UNEP and WMO. Cambridge university press. 1996.

FIG. 2.2 The greenhouse effect (Image courtesy of UNEP/GRID-Arendal Maps and Graphic Library (2002). Retrieved from http://maps.grida.no/go/graphic/greenhouse-effect)

Forcings and Feedback

In climate change science some other specific terms come with the turf. "Forcings" and "feedback" are perhaps the most common of these. We can have both positive and negative forms of both, depending on whether the net result is an addition or a loss of temperature. Generally speaking, positive forcings are those primary factors that initially add to the total energy equation, opposed to the secondary effects of positive feedback as existing matter re-radiates absorbed energy to other matter. Positive feedback is thus a secondary process, whereby matter that has been warmed via the primary forcing agent then radiates energy to other surrounding matter, only to return it back to the original matter in a circular fashion. This process compounds the warming effect.

The subject is somewhat diffusely defined, since many feedback gases, such as water vapor, also have forcing properties. Thus, gaining a precise understanding of the differences between both types of effects may be confusing. The IPCC has defined a forcing, technically, as:

> a measure of the influence a factor has in altering the balance of incoming and outgoing energy in the Earth-atmosphere system and is an index of the importance of the factor as a potential climate change mechanism .... Radiative forcing values are for changes relative to pre-industrial conditions defined at 1750 and are expressed in watts per square meter (W/m^2).

Additionally, the IPCC's definition of feedback is:

> an interaction mechanism between processes in the climate system ... when the result of an initial process triggers changes in a second process that in turn influences the initial one. A positive feedback intensifies the original process, and a negative feedback reduces it.

Thus, in the loosest and simplest terms, radiant responses from water vapor, already part and parcel of the natural atmospheric environment, are considered overall as feedback, even when the primary effect may be partially that of a forcing. Depending on climatic conditions at the time, water vapor comes and goes rapidly. Thus, it is a highly variable and volatile component, unlike carbon dioxide (CO_2), and this goes to the crux as to why it is treated differently. Similarly, one might argue that although CO_2 exists in a natural state throughout the world (most CO_2 is, in fact, natural) and has been warmed partly by water vapor, that it should be considered feedback instead.

Water Vapor *Not* a Forcing Agent?

A German website, Allianz.com, declares that although water vapor is not a forcing agent, it amplifies existing warming.

Regardless of this confusing description or how one categorizes it, water vapor is responsible for 21 K of the total (33 K) natural greenhouse warming effect. Thus, Allianz's position illustrates the difficulties of separating forcings from feedback.

One could also argue that many activities of modern humans, such as the combustion of fossil fuels, lead to more water vapor in the atmosphere, and thus, it should be more accurately categorized as a forcing agent. It does absorb infrared radiation from Earth, albeit at different frequencies to CO_2, and subsequently re-radiates it to other adjacent matter, which may result in the formation of more water vapor! So perhaps the easiest way to consider the primary difference between them really is to consider just the time of residence in the atmosphere. Since CO_2 resides in the atmosphere for a much longer period of time (more about this later), and mostly is increased by manmade emissions and activities (as opposed to by naturally occurring processes), it is considered a forcing agent.

Needless to say, the confusion may remain for many readers! The distinction between them still remains a fine line. For anyone wanting a more detailed description of these somewhat esoteric distinctions, an excellent and very readable analysis may be found online at RealClimate.org [1].

Beyond the warmth that we depend on for life itself, we need to know, therefore, how the various natural and anthropogenic forcings and feedbacks

- Interact with each other.
- Counteract each other.
- May result from the presence of another, and to what degree.
- Are counterbalanced, decreased or enhanced by aerosols (atmospheric particulates, solid or droplet).
- Are affected by varying degrees of Earth's albedo (the diffuse reflective properties of incident radiation by such a body of matter).

We could expand this short list almost limitlessly. More to the point, though, have we yet reached runaway status, whereby a chain of unstoppable forcings and feedbacks will take place in a cascade? Conversely, is there anything we can do – or should do – to mitigate these effects? Perhaps most significantly, is it all our fault, or at least partly so?

Black Bodies in Space

In addition to the fundamental mechanisms of Earth's "greenhouse," we need also to understand something of the physical process whereby matter absorbs electromagnetic energy and re-radiates it at different wavelengths. This process is central to the principles involved in maintaining Earth's climate.

Let us consider first what is termed a "black body" (something existing only in the purest theoretical sense) – a hypothetical object that absorbs all electromagnetic radiation that falls upon it and re-radiates all of it as thermal radiation. At low temperatures, such objects, were they to exist, would appear perfectly black – invisible, in fact. Increasing temperatures of re-radiated thermal energy would cause spectral emissions ranging from red to white, with pure white objects (the hottest) emitting a substantial portion of energy in the ultraviolet part of the spectrum, depending on the level of extreme temperature at the upper end.

Let us imagine a black body of identical size, round shape and spectral makeup to the Sun. If it were to re-radiate all of the energy that it absorbs, what we would see would appear much the same as the Sun, having virtually the identical radiant power and spectrum. This scenario makes for an interesting and remarkably close comparison [2], also having a temperature of approximately 6,000°C, and would exhibit the pure white of the upper end of the spectrum, including ultraviolet wavelengths. Although the Sun itself isn't a black body, of course, since it generates its own radiation internally, the visual comparison is instructive, nevertheless.

For another example in the Solar System, let us look at Earth. Since it re-radiates a good portion of the energy it receives from the Sun in the form of low frequency invisible infrared waves, Earth is acting like a partial black body at the lower, cool end of the electromagnetic spectrum. It is partial because some of the Sun's radiation (in all wavelengths) is reflected back to space.

Initially, about 30% of incoming solar energy (including light wavelengths) is reflected (as well as scattered) directly back to space from the top of the atmosphere, by layers of clouds and Earth itself, which is what renders our world visible! What remains is absorbed by Earth and its atmosphere and re-radiated at low frequency infrared wavelengths into the surrounding atmosphere, land, oceans, and space.

Instinctively, even if the term is new to us, we already know about the concept of black bodies, since we think regularly in terms of "red-hot" and "white-hot" objects. Even if we have never considered it, though, we realize instinctively that with increasing temperatures objects re-emit absorbed energy at higher wavelengths. We can envisage already how black bodies at low temperatures would appear by thinking of familiar things such as optical blacking, soot, black felt, velvet and the like, even though the latter examples would hardly survive any radical increasing temperature! The opposite, of course, to a black body would be a mirror-like object and other very shiny (reflective) surfaces that absorb very little energy.

Although electromagnetic energy cannot be destroyed, it can be transformed into energies of different wavelengths. The portion of solar energy that is intercepted by Earth must be accounted for in an energy equation. The total incoming radiation is measured at 342 watts per square meter (abr. 342 W/m²). Whether absorbed and retransmitted at lower wavelengths or reflected back into space, it represents a simple concept, although it is very complex in practice. Regardless, it all must add up and be accounted for.

Continuing to look at Earth as a partial black body, approximately 16% of incoming radiation is absorbed by the atmosphere and clouds, and the remaining 54% or so by the surface. Of the remaining 30%, that which is reflected is dependent on the specific planetary albedo, as well as the reflectivity and amount of cloud. The fraction that is scattered is dependent on the molecules of the atmosphere (a function of Raleigh scattering) and aerosols, which may even include water droplets in the atmosphere (a function of the anomalous diffraction theory).

The 1971 NASA study called "Earth Albedo and Emitted Radiation" [3], summarized its conclusions with the following:

- Reflectance tends to increase with increasing solar angle.
- Continental areas have higher albedo values than ocean areas.
- Albedo increases with latitude due in part to the decreasing solar elevation angle, snow and ice near the poles, and increased cloud cover associated with large scale weather activity.
- Regions of dense cloud cover have higher albedo values.
- Albedo values for any region vary seasonally, primarily because of changes in cloudiness, vegetation and snow and ice cover (Fig. 2.3).

All in all, it is daunting to produce any realistic computation of the combined effect.

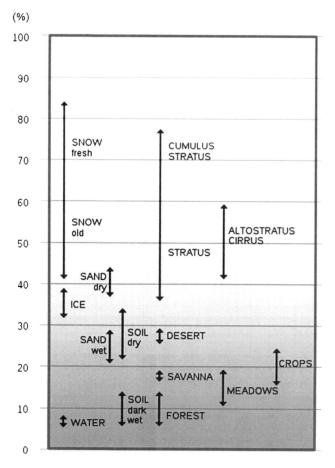

FIG. 2.3 Earth's albedo, in percentages according to surface and atmospheric conditions (Graphic by Hannes Grobe)

The albedos of different terrestrial surfaces, vegetation or clouds vary greatly. More specifically, desert and arid surfaces are reflective to some degree, but snow is almost entirely so; cold water is moderately reflective, warm water much less. The densest, lowest clouds tend to be the most reflective (by as much as 75 % in some cases) and high cirrus clouds the least. However, the continuing variability of clouds as a whole makes exact measurements of their reflective properties among the most problematic components in the entire study of climate change.

The forcing and feedback effects of such large and variable amounts of potential cover and reflectivity are large, but the process may be a lot more complex than previously realized, as

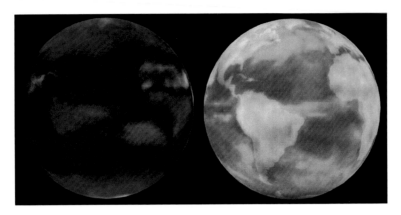

FIG. 2.4 The radiative properties of Earth. (*Left*) Heat given off by Earth's surface and atmosphere and radiated to space. (*Right*) Sunlight reflected out into space by land, oceans, clouds and aerosols (Data provided by the Atmospheric Sciences Data Center and CERES Science Team at NASA Langley Research Center; Image courtesy of Todd Bridgman, NASA Goddard Space Flight Center Scientific Visualization Studio)

shown in the study by Timothy Andrews and Piers M. Forster [4]. Although Earth's albedo determines how incident radiation is reflected, the remainder is absorbed, to be re-radiated at low frequencies. This is clearly demonstrated in Fig. 2.4.

Theoretically, at least by calculation, the direct effects of the Sun should warm Earth from absolute zero (0 K, or –273.15°C) to 270 K. Because Earth does not absorb all the incident energy, its actual direct warming is estimated to be only approximately 254 K. Again, by calculation, scientists have deduced the total warming due to greenhouse gases to be approximately 33 K, to give a total average global temperature of approximately 287 K, or 13.85°C.

The Atmosphere and Interactions with Solar Radiation

During the forcing process, radiation absorbed by Earth and its atmosphere is re-radiated in all directions to adjacent molecules. In greenhouse warming, it is mostly the absorption by the constituent gases of the atmosphere that creates the effect. In turn the process continues as these molecules re-radiate to other molecules during the feedback process.

CO_2 would seem to be the only changing factor measured with certainty over the past 40 years. Changing concentrations of, say, water vapor have been much harder to determine. Depending on the source for exact amounts, the total amount of CO_2 in the atmosphere today stands about ± 100 ppmv higher than it did before industrial times, or approximately 32%. Different estimates shade this amount up or down a little, but it is a good reference point for our purposes.

Perhaps surprisingly, most of the minor constituents of the atmosphere exist essentially in uniform concentrations throughout its volume, regardless of altitude; carbon dioxide occupies surprisingly little of it at ± 0.0388%, although next to water vapor it is the leading greenhouse gas. Ozone concentrations and water vapor, however, unlike all the trace greenhouse gases, are *not* evenly distributed throughout the atmosphere. They are both variable in amount, producing a less predictable greenhouse effect.

Although water vapor is an invisible greenhouse gas, upon condensing into clouds it becomes more a reflective cooling component than an absorptive one – a negative feedback. Low atmospheric ozone is produced as a result of solar photochemical reactions with vehicle emissions, a direct byproduct of fossil fuel burning. It is common in urban areas, acting as a forcing agent. However, ozone is present higher in the stratosphere, where it acts reflectively as a negative feedback! Thus, ozone depletion from chlorofluorocarbons (CFC's) in this part of the atmosphere results in additional warming potential. It is easy to see why the total effects of ozone and water vapor remain something of a wildcard (Fig. 2.5).

We should be sure to note again the actual proportion of CO_2 in the atmosphere. Since its concentration is 380 ppmv (precise readings depend on the source), or only 0.0380% of the total atmosphere, we are certainly far from suffocating in this gas. Such popular misconceptions should not, however, lead us to dismiss increasing concentrations lightly. In 1750 CO_2 concentrations were approximately 0.0288 ppmv [5]. Despite what appears to be a minimal concentration in the atmosphere – both in 1750 and today – the effect of carbon dioxide must not be discounted, because regardless of the amount of warming it causes, it has fairly strong forcing capabilities compared to many other atmospheric gases. Increased concentrations also persist in the atmosphere for a substantial amount of time, and lie at the very heart of the present controversy.

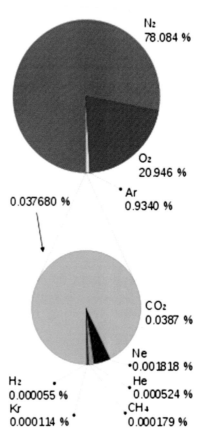

FIG. 2.5 Proportions of gases of the atmosphere (not including water vapor) (Graphic by Mysid)

More specifically, we can now analyze energy being re-radiated from Earth over any time frame, and it is clear how events in the climate can influence it. The NASA graph (Fig. 2.6) shows variations in Earth's long-wave (infrared) radiation from tropical latitudes averaged over more than two decades. Variations in its amplitude can clearly be seen, caused by different events at or near the surface. They are most readily apparent, due to atmospheric aerosols, following volcanic eruptions, as well as during El Nino years. The former may be identified by strong dips during volcanic eruptions, and distinctive radiative spikes during the latter. Aerosols not only prevent a portion of solar radiation from reaching Earth (contributing to an overall cooling effect) but also leaving it (adding to warming). Overall, however, their effect is of a negative forcing.

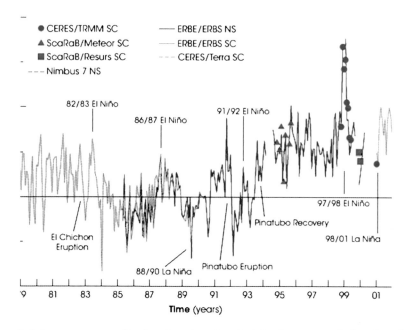

FIG. 2.6 Long-wave radiation leaving Earth (*the horizontal dividing line representing the mid-point between reductions in radiated infrared energy below the line up to 0 W/m²*), and increases in radiated energy (above the line up to 10 W/m²) (Graph courtesy of NASA)

The sharp upwards spike of 1997/1998 is remarkable, not only due to its strength but also because after this event there was a marked step upwards in global average temperatures of 0.25°C that has persisted since in the baseline average global temperature [6]. Thus, claims of *gradual* temperature increases in step with CO_2 increases over the period do not appear to be substantiated. It will be interesting to see if the substantial Icelandic and Indonesian volcanic eruptions of 2010 can be tied to any decreases in radiation or temperature. It has already been remarked by some, however, that climate models projected far greater temperature declines after sizeable volcanic eruptions over the past half-century than actually were observed.

Although the IPCC has stated that the long cooling trend from around 1940 into the 1970s can be largely attributed to the total of aerosols from all origins (including volcanoes), this also appears to be hard to reconcile with the volcanic record and assessments of present warming trends. However, even if we accept that in theory volcanic aerosols could cool the climate, it might be seen

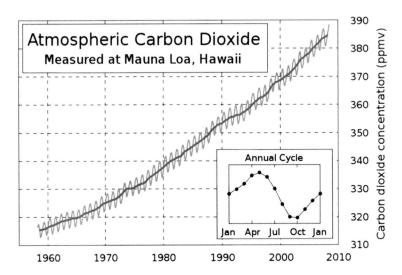

FIG. 2.7 The rise in atmospheric CO_2 since 1960 (Graph by Sémhur)

as unlikely that any scientist would suggest the possible release of aerosol particulates into the atmosphere as a means to offset present day warming, even less to actually recommend such actions. Remarkably, though, this is not the case at all.

A program in the United Kingdom with the acronym SPICE (Stratospheric Particle Injection for Climate Engineering) has proposed to do just that! Although recently placed on hold for some months, it promises to test the concept with fine water-sprayed droplets from a high altitude balloon, with the hope of graduating to full-scale sulfur-type particle injection later. The researchers believe that up to 2°C cooling could be achieved artificially, but naturally, many geopolitical and legal factors have yet to be ironed out.

With the majority opinion within the IPCC that present-day climate change is human induced, apparently it is no longer seeking other underlying causes. However, its position regarding the period between 1940 and 1970 remains a sticking point, even though climate models have been adapted to reflect this scenario. The gradual, virtually monotonic, upward curve in CO_2 over the past 100 years or so bears no resemblance to the wildly uneven global temperature footprint at all, even proportionately. Other than showing a total rise in both from the beginning of the century to the end, there are few other visible parallels (Fig. 2.7).

Anthropogenic Greenhouse Gases

Despite the apparent clarity of the various AGW greenhouse gas contributions shown in the next graphic (Fig. 2.8), greater understanding is needed of variables, including:

- The enormous complexity of Earth's changing albedo.
- Changes in water vapor.
- Possible yet unrecognized consequences.
- Possible repercussions of changes in the output of any gas, or its proportion.
- The precise role that increasing concentrations of carbon dioxide and other greenhouse gases might play.

Such factors remain undetermined, especially since we can only speculate on most of them. We can see in this chart

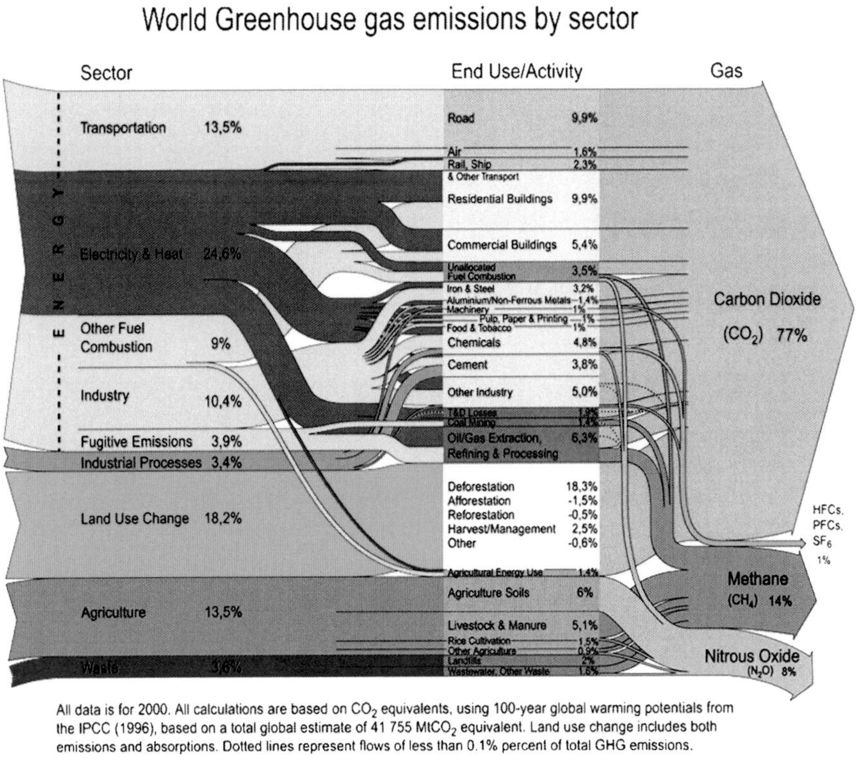

World Greenhouse gas emissions by sector

Sector		End Use/Activity		Gas	
Transportation	13,5%	Road	9,9%		
		Air	1,6%		
		Rail, Ship & Other Transport	2,3%		
		Residential Buildings	9,9%		
Electricity & Heat	24,6%	Commercial Buildings	5,4%		
		Unallocated Fuel Combustion	3,5%		
		Iron & Steel	3,2%	Carbon Dioxide	
		Aluminium/Non-Ferrous Metals	1,4%	(CO₂) 77%	
Other Fuel Combustion	9%	Machinery	1%		
		Pulp, Paper & Printing	1%		
		Food & Tobacco	1%		
		Chemicals	4,8%		
Industry	10,4%	Cement	3,8%		
		Other Industry	5,0%		
		T&D Losses	1,2%		
Fugitive Emissions	3,9%	Coal Mining	1,3%		
Industrial Processes	3,4%	Oil/Gas Extraction, Refining & Processing	6,3%		
		Deforestation	18,3%		
		Afforestation	-1,5%	HFCs,	
Land Use Change	18,2%	Reforestation	-0,5%	PFCs,	
		Harvest/Management	2,5%	SF₆	
		Other	-0,6%	1%	
		Agricultural Energy Use	1,4%		
		Agriculture Soils	6%	Methane	
Agriculture	13,5%	Livestock & Manure	5,1%	(CH₄) 14%	
		Rice Cultivation	1,5%		
		Other Agriculture	0,9%		
Waste	3,6%	Landfills	2%	Nitrous Oxide	
		Wastewater, Other Waste	1,6%	(N₂O) 8%	

All data is for 2000. All calculations are based on CO_2 equivalents, using 100-year global warming potentials from the IPCC (1996), based on a total global estimate of 41 755 $MtCO_2$ equivalent. Land use change includes both emissions and absorptions. Dotted lines represent flows of less than 0.1% percent of total GHG emissions.

FIG. 2.8 Greenhouse gases due to humans (anthropogenic gases) (Image courtesy of Emmanuelle Bournay, UNEP/GRID-Arendal; available at: http://maps.grida.no/go/graphic/world-greenhouse-gas-emissions-by-sector)

the breakdown of things as they stand at the present time but should be cognizant of the uncertainties dictated by evolving circumstances.

On AGW CO_2

From the NRDC (National Resources Defense Council):

"Carbon dioxide and other air pollution that is collecting in the atmosphere like a thickening blanket, trapping the sun's heat and causing the planet to warm up. Coal-burning power plants are the largest U.S. source of carbon dioxide pollution – they produce 2.5 billion tons every year. Automobiles, the second largest source, create nearly 1.5 billion tons of CO_2 annually."

And:

How can we cut global warming pollution?
A: It's simple: By reducing pollution from vehicles and power plants.

It is hard to forgive so many scientific misconceptions and transgressions within just one sentence, since

- Anthropogenic carbon dioxide is not a pollutant in the normal sense of the term, despite the absence of the 14C signature in naturally occurring CO_2.
- It does not *collect* in the atmosphere like a thickening blanket, since it is essentially evenly distributed throughout the atmosphere relative to volume.
- CO_2 does not *trap* the Sun's heat.
- CO_2 has not yet been conclusively demonstrated to *cause the planet to warm up.*
- Automobiles are not the second largest source of CO_2. Look again, even at the IPCC's own analysis (Fig. 2.8), in which *all* transportation accounts for just 13.5% of all human-produced CO_2, of which automobiles account for just 9.9%.

It is clear that increasing temperatures will also result in increasing water vapor. It is generally considered that the natural water vapor balance, as well as its effect, can be rapidly affected by its response to increases in the amounts of any other greenhouse gas emissions. However, there is a continuing controversy about exactly how much warming is caused by increased greenhouse forcing gases, and thus how much water vapor results from them. This is in addition to the measurement of extra forcing and feedback that might result from the increased water vapor itself in the atmosphere.

We should also bear in mind that although these gases, together with man's own additional contributions of CFC's may be seen in Fig. 2.8, water vapor accounts for almost all of the atmosphere's greenhouse gas, as well as the warming, followed by other gases, which trail far behind. By many calculations (see later), of the 33°C total natural greenhouse warming of Earth, that from water vapor amounts to at least 95% of it. This is exclusive to the present climate change discussion, since we are only referring to Earth's natural greenhouse. There is no universally accepted ratio, however.

Thus we can readily see that CO_2 and other trace component gases fall into controversial territory. Of the group of remaining greenhouse gases (not including water vapor), CO_2, being a moderately strong absorber and source of re-radiation, accounts for at least 77% of the warming. Despite its relatively minute proportion by mass compared to atmospheric water vapor, it is a more significant greenhouse gas, and responsible for the higher ratio of the warming itself than might be expected.

The degree that carbon dioxide versus water vapor is responsible for warming the lower atmosphere has been the ongoing subject of dispute throughout all of the present climate controversy. Depending on the source, the numbers are surprisingly different. When one logically thinks through some of the higher estimates for CO_2 warming, they might seem unnecessarily alarmist. Although they could be accurate, it is hard to find coherent explanations for these conclusions – at least in searches one can readily conduct oneself. This does nothing to calm the controversy.

Adding It Up: The CO_2 Equation

T. J. Nelson, a research scientist with a Ph.D. in biophysics, works in the field of biomedical research. He has found a passion for climate issues, although we should be aware that his self-published views on climate, while solidly conceived, do not constitute "peer-reviewed" study. However, they seem worthy of our attention, particularly with the dearth of readily available similar analyses. In his online article "Cold Facts on Global Warming" [7], Nelson shows how he calculated by formula the specific warming due to CO_2 to be between 4.2% and 8.4% of the total greenhouse warming effect.

On the face of it at least, this puts into perspective what we are actually dealing with. Other sources, including the IPCC, put the influence of CO_2 at a much higher number; for example, Wikipedia lists its warming influence at 9–26%, and water vapor at 36–72%. Thus we have possible total temperature increases from 1750 until the present, and those caused by greenhouse gases, (excluding water vapor) ranging from 1.39°C (Nelson), to 6.4°C (IPCC AR4), and to 8.6°C (Wikipedia). All in all, this allows for quite a wide berth in the margin of error. Of the trace gases, according to Nelson, carbon dioxide warming counts for 84% of the total.

Most sources inform us that the total warming on Earth since 1900 is about 1°C; some shade it up or down a little. Regardless, the lack of agreement on the effects of CO_2 on temperature causes us considerable difficulties if we try to do the math. The answers produce seemingly irreconcilable conclusions based on previous atmospheric concentrations. If we take the upper figure for the effect of CO_2 (i.e., that of Wikipedia), none of the natural ratios of warming relative to those of recent times seems to make any sense at all.

Nelson does take issue with a "pro-consensus" website, www. realclimate.org, and its estimate that actual CO_2 warming is closer to five times his final projection of ~5% of the total (a number based on what he terms "most credible sources"). In the equation he presents, the IPCC's projection simply is not possible, given past history and the makeup of existing atmospheric warming. He argues that if we use the projected inputs and responses of IPCC models to retroactively calculate present-day temperatures, they should have reached what he terms "preposterous heights." Thus, in his view, such models have been made to fit what has actually taken place.

Since we know how some of the variables in Earth's complex system affect climate, nothing can be taken at face value. However, perhaps the most interesting and provocative point made in Nelson's paper is his analysis that shows that the atmosphere already is sufficiently laden with CO_2 (and is effectively transparent to incoming solar infrared radiation of higher wavelengths) to have absorbed most of the low wavelength infrared radiation being re-emitted from Earth's surface. Accordingly, additional amounts of the gas would generate differences from the added concentrations only. Therefore, these would not compound the warming in a multiplying manner. Any increases would follow a logarithmic curve, whereby the amount of heating already produced by the total would only be added to incrementally. Even if atmospheric CO_2 were doubled (to 760 ppmv!), the worst-case scenario he projects would be a temperature increase only of between 1.02°C and 1.85°C. Similar temperature increments would be the case with *each subsequent doubling.*

At the present rate of increase, Nelson deduces it would take a couple of centuries to reach those levels. Additionally, increases or decreases in anthropogenic production CO_2 would be immediately felt, since the atmosphere cannot store it for long (5–10 years). This figure is in direct conflict with other claims (50–200 years, and more!) that we have heard.

Debate continues over the "shelf life" of CO_2 and other greenhouse gases in the atmosphere. Clearly, this is a significant part of any argument, since projections of a lengthy existence of anthropogenic greenhouse gases residing in the atmosphere color any discussion about their potential to affect climate. If we accept the low estimate of only 5–10 years, a remedy for any proven ill effects of the gases on climate ought to be the most straightforward to find.

CO$_2$ Residence Time

The website skepticalscience.com has claimed that excess human emissions of CO_2 will remain resident in the atmosphere for over 100 years.

> Again, such approximations of fact do this website grave injustice, not only in the simplistic and imprecise picture painted but by implication that only *human* CO_2 emissions have a long residence time.

Of course, increased temperatures would increase water vapor, which, being a greenhouse gas, would also add some amount to the warming; Nelson does not appear to build this aspect into his math. He does, however, take us through a projected scenario that includes Earth's responses to increased warming (its albedo from increasing or decreasing clouds and snow cover, etc.) and once again demonstrates, by his carefully laid-out logic, that total warming is still well within the basic 1.85°C range – his maximum estimate. Even with numerous other potential and actual factors, his logic seems consistent, providing some very insightful food for thought, whether or not one regards his view as being the last word.

In contrast, if we accept the IPCC's position, things look a lot more serious. Should anthropogenic greenhouse gases be the root of modern climate change, the IPCC maintains that, even if emissions stopped completely: "About 50% of a CO_2 increase will be removed from the atmosphere within 30 years, and a further 30% will be removed within a few centuries. The remaining 20% may stay in the atmosphere for many thousands of years," in which case, it is already too late to remedy some damage that has already been done.

Also, the atmospheric lifetime of water vapor is quite transitory at only about 10 days, something that is not disputed. Thus, it is primarily CO_2 that is our concern. Apparently, the fear exists that this gas may tip the scales of compounding greenhouse warming conditions, leading to a runaway crisis. A study, by Arindam Samanta et al. [8], concurred with the more pessimistic findings. Considering the limited warming potential of these gases, as well as far lesser concentrations, not everyone agrees with this line of reasoning. It would probably be fair to state that most researchers believe that the true warming potential of any gaseous compound depends most critically upon its total ability to absorb and re-radiate heat efficiently, as well as its residence time in the atmosphere. Settling this argument would do much to resolve some key disputes behind the AGW theories.

More typical of articles and papers in defense of the theory of carbon dioxide driven warming is "Atmospheric CO_2: Principal Control Knob Governing Earth's Temperature," by Andrew A. Lacis et al. [9]. Assessments easily drawn from articles such as these only serve to keep the controversy alive. The article states that CO_2 accounts for 20% of the warming from greenhouse gases. By that line of reasoning, most casual readers will deduce different outcomes on a number of levels:

- If the total warming prior to 1850 accounts for thus 20% of 33°C (the calculated average natural global warming above simple solar warming), it would equate with a total warming effect from natural CO_2 to have been 6.6°C.
- By the same line of reasoning, should not the 92 ppmv of CO_2 that has been added since then have produced an additional warming of 2.17°C: $6.6 - 280 \times 92 = 2.17°C$, and not just the ~0.8% recorded? However, the 0.8°C temperature increase since 1850, give or take a few fractions of degrees by various authorities, is hardly in dispute.
- Perhaps the implication meant was that 20% of the actual measured warming is the result of the approximately 92 ppmv of CO_2 added since 1850. However, if this should be the case, 0.16° of warming would be all that the added CO_2 is directly responsible for! Since with such a result it could hardly be claimed that there is an environmental problem, this cannot be the authors' meaning.
- Or, are the authors possibly referring to the feedback warming effect beyond the primary greenhouse foundation to be at a different ratio of 20%? If so, they certainly do nothing to dispel the notion that radical warming of this magnitude could result from just small CO_2 increases.

We shouldn't have to guess what the article means, and ultimately, whose facts to believe. Despite coming from such eminent scientists, the article illustrates exactly the kind of confusing positions that have become all too common in discussions about climate. How could such wide disparities in estimates have occurred? Even the IPCC itself seems to consider that a doubling of CO_2 by itself would only increase temperatures by 1°C, despite the different stated percentages they have provided of its warming effect.

The IPCC also concedes that 0.12°C of the total increase since 1900 can be attributed to what is known as the Urban Heat Island Effect, where zones of densely populated areas have been theorized by some to produce misleadingly high readings. It becomes ever clearer that the science of climate change is so complicated, laden with theories, studies, projections, statistics, findings, contradictions and disagreements, that any newcomer to the topic will likely reach sensory overload very quickly. Regardless, it further demonstrates how much remains to be settled, and how unsupported – or unexplained – statistics continually lead us into trouble.

> In contrast, the IPCC summed up this effect differently: "However, over the Northern Hemisphere land areas where urban heat islands are most apparent, both the trends of lower-tropospheric temperature and surface air temperature show no significant differences. In fact, the lower-tropospheric temperatures warm at a slightly greater rate over North America (about 0.28°C/decade using satellite data) than do the surface temperatures (0.27°C/decade), although again the difference is not statistically significant."

> TAR, 2001

It is interesting to note that statistician Thayer Watkins of San José State University Department of Economics deduced virtually the same degree of inaccuracies in IPCC projections of CO_2 warming – apparently by coincidence in his own independent and carefully constructed study [10]. However, once again, as a representation of his views only, it is not a peer-reviewed paper. His conclusion that the IPCC's projection is ~2.5 times reality hits about same note as we have just illustrated. More controversial, though, will be his conclusion that the IPCC uses such projections to "generate scary projections," although he is not the first to make such claims.

Among other scientists who have taken up the issue of carbon dioxide, Jeffrey A. Glassman (an applied physicist and engineer), in another independent paper posted on his own website [11], approaches the issue a little differently, although it is closely related to that of Nelson's approach. Using the same basic principles of physics outlined by Nelson, Glassman compares them to the record taken from the Vostok ice core. This core was found and extracted in 1988 and was the deepest ice core ever found and recovered (3,623 m), extending back through history some 400,000 years. This is an excellent proxy record, providing invaluable information on past climate cycles, along with the trace gas composition of the atmosphere. In this case, especially, the record of atmospheric carbon dioxide concentration appears clear.

Glassman also discusses oceanic respiration (outgassing) of CO_2, that is reabsorbed back into cold ocean waters and transported down into the deepest, most highly saturated regions. Increases in atmospheric CO_2 have always accompanied increasing temperatures, but delayed in time. As Nelson similarly observed, if global temperatures had followed the projected warming that would accompany such increased CO_2 concentrations as in the ice cores, the consequences already would have been disastrous. Glassman's final take, however, is a little different. He concludes that increased CO_2 must have always been reabsorbed into the oceans before such events could take place, an interesting hypothesis. He also considers increased CO_2 quite likely to be the result of warming by some yet undetermined process of outgassing and not the other way around.

Greenhouse Gases as a Shield

The components of the atmosphere shield us from some of the Sun's rays, including those that are especially harmful. We can also see (Fig. 2.9) that the specific wavelengths of solar radiation absorbed by the various atmospheric components differ quite significantly. Thus, the absorptive function of one does not necessarily overlap, or necessarily enhance (or detract), that of another. This is another reason that the equation is so complex. We can see from this graph (averaged for different regions) how the various atmospheric components determine the degree to which different wavelengths of the solar spectrum are absorbed.

Different trace gases also absorb energy and re-radiate infrared wavelengths most efficiently at specific temperatures. For example, carbon dioxide absorbs and radiates heat to the maximum extent at low temperatures in the region of 220 K. The altitude this corresponds to is 40–50 km. At lower altitudes, increasingly warm

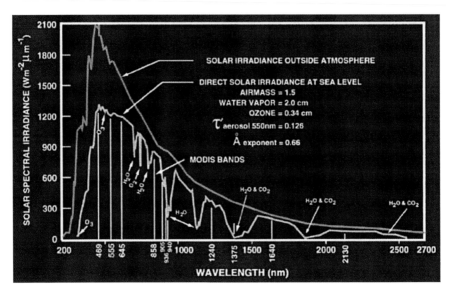

FIG. 2.9 Atmospheric transmissions (Image courtesy of NASA)

temperatures make that part of the atmosphere less able to radiate energy back to space. In order to maintain equilibrium of energy absorbed versus energy lost, more infrared energy will be transmitted back to space from the altitude where the temperature for re-radiation is most efficient for that gas. Increasingly cold temperatures at high altitudes will be the result, and warmer ones will be closer to the surface – a chain reaction of forcing and feedback. Because the total incoming energy is effectively measurable, many researchers consider that confirmation of cooling temperatures in the upper atmosphere would be significant in showing the influences of anthropogenic gases. There is some evidence – admittedly not conclusive or overwhelming – that this could indeed be taking place (see Chap. 4, Fig. 3.19?).

Carbon Sinks

Closely related to this topic, and indeed part of it, are the natural "sinks" of Earth, and their ability to mitigate the increased concentrations of heat and CO_2 by absorption. By far the largest reserves of carbon dioxide reside in the oceans and landmasses. An interesting history by Richard Mackay about the work of pioneer Rhodes Fairbridge raised the issue of CO_2 outgassing from the oceans – the opposite action to absorption. He concluded that increased outgassed CO_2 from Earth's "sinks" might have contributed a portion of the increased temperatures since 1850 [12]. However, Mackay did not speculate on how much of the increase can be ascribed to this natural process, presumably feeling this was beyond the realm of the paper, perhaps even his particular expertise despite his considerable credentials.

The mechanics of outgassing are simple. Because increasing temperatures reduce the solubility of CO_2 in water especially (or in soils), this causes increased natural reductions of CO_2. However, even though the added CO_2 in the atmosphere is very small, it is again the increasing concentrations that have come under the microscope, because most researchers do not believe they have been induced naturally.

Although increased anthropogenic emissions may be responsible for the majority of CO_2 increases, there is an ongoing recycling of CO_2 by the oceans and land. Some have questioned whether the observed changes in atmospheric concentrations are due to some not yet understood processes [13] but many "skeptics" have argued that increased temperatures are the sole cause of increased carbon dioxide, especially in the last 100 years – that greenhouse gases are the outcome of natural warming variations. However, this does not appear to be credible, at least according to the laws of physics and chemistry Nelson outlined with such clarity. The increased CO_2 concentrations under discussion are far greater than could have been released by the oceans following the recorded temperature rise, in which the amount of outgassed CO_2 would be a mere 4.4 ppmv following an increase of 0.6°C. For the actual average temperature increases usually mentioned since 1900, this would amount to about 5.9 ppmv, a far cry from the amount we have actually seen.

However, it should be noted that Nelson says nothing about the involvement of landmasses, which also have large reserves of CO_2, especially from ongoing or old decaying vegetation from deforestation. The omission seems perhaps sound, since Earth's

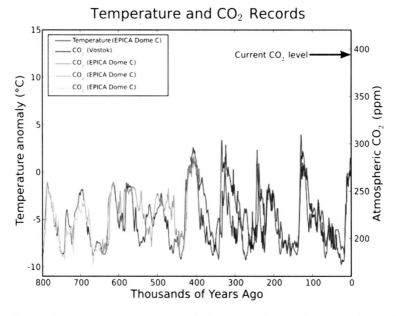

FIG. 2.10 Historic temperatures and CO$_2$ correlation (Graphic by Leland McInnes)

landmasses represent a far smaller CO$_2$ sink than the oceans. However, the present day clearing of forests across large continental landmasses, with the subsequent increased volumes of carbon emissions, may play a larger role than Nelson's paper suggests. This is even after we bear in mind that landmasses would still encompass hugely less CO$_2$ than the oceans.

Only in the last 100 years or so do we find no obvious correlation between increasing concentrations of CO$_2$ and the historic record. Prior to this, the ratio of carbon dioxide relative to temperature can be seen to be consistently predictable, where substantial and proportional temperature increases apparently have always followed elevated CO$_2$ atmospheric concentrations (Fig. 2.10). The ~12°C swings in temperature that follow CO$_2$ concentrations are striking, but do not appear to correspond to:

- Possible amounts of outgassing from such temperatures.
- Any existing theory of comparable warming resulting from CO$_2$ increases.

Global Flows of Carbon

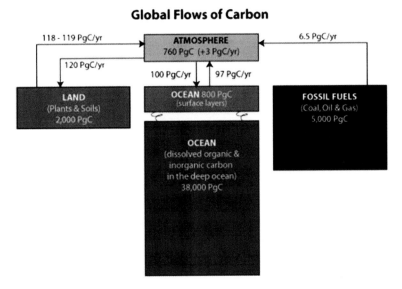

FIG. **2.11** The carbon cycle (Graphic courtesy of NASA)

Therefore, it is reasonable to assume that both of these synchronized swings must have been caused by one or more other stimuli external to any direct relationship. Measurements from hundreds of millions of years ago reveal CO_2 concentrations 10–20 times present-day levels, contrary to claims by some that they have never been as high as they are now. Some might reference this to suggest that similar undetermined forces might have produced the rise in CO_2 seen during the last 100 years.

Regardless, some researchers have suggested that the naturally occurring carbon sinks can more than absorb all of the increasing greenhouse gases, as well as any increasing temperatures. Thus, presumably these sinks could have always acted as long-term regulators in some way.

A study by Joseph G. Canadell et al. [14] rated the amount of natural absorption of anthropogenic CO_2 into natural sinks at 57%, which is not inconsiderable. This seems to indicate that CO_2 lifespan in the atmosphere could indeed be short, despite many claims to the contrary. It also tells us that the maximum we would need to reduce emissions by would be just 43% of the total. In other words, according to the time at which the chart (Fig. 2.11) was made, and using the most superficial estimate, less than

2.8 GT a year – assuming this could finally be agreed upon by all concerned as being a necessity and solution.

If such absorption does not produce serious changes to the chemical composition of the oceans alone (a "sink" so large, at 38,000 GT, it is hard to conceive that in the short term an annual increase of 2.8 GT could statistically alter it in a meaningful way), the fact remains that, according to Canadell et al., just 2.8 GT represents the probable *total amount* that CO_2 levels are out of balance. It is argued that the cumulative effect will build over time, but with amounts such as these, perhaps there will yet be a way found to equalize atmospheric conditions to pre-1750 levels, again if that were to be determined as the remedy to additional warming.

If we are looking to know the length of time it would take to reach equilibrium again (of both CO_2 levels and temperature), this remains undetermined, so the conundrum continues. However, we do know that the oceans, and biosphere, have been absorbing large quantities of anthropogenic CO_2 all along, and that levels of the gas in the atmosphere would have been hugely greater than they are at this time. It could be argued that the sinks have already been tested under extreme conditions through history, and proven an effective shield against climate change, but this is pure conjecture at this stage.

A bigger question is whether corrective actions would produce any effect at all, according to the so-called "skeptical" scientists, since they do not accept the AGW theory in the first place. Regardless, if we do accept them, we know that without anthropogenic contributions, at 1750 temperature levels apparently the balance was reasonably stable.

The process of ocean CO_2 storage is interesting. Acidic carbon dioxide oxide gas (CO_2) dissolves in seawater to form carbonate and bicarbonate ions in reactions with calcium rock deposits on the ocean floor, resulting in reduced alkalinity (of 0.1 units, per the IPCC) and greater ocean acidity. Because the oceans are warmest near the surface, they are less able to hold as much CO_2, which is thus increasingly outgassed into the atmosphere, according to the degree of ocean warmth.

The denser, more acidic waters drop to ever greater depths, producing a very highly carbon saturated liquid of ever colder temperatures. Of the total Earth reserves, anthropogenic annual CO_2 contributions amount to approximately 6.5 GT (see Fig. 2.11 [15]; *GT* is termed in the NASA graphic as *Pg C*). This is quite a small percentage of the estimated total that is exchanged back and forth in the biosphere. The total CO_2 in play is close to 1,000 GT per year.

A good guide to the amounts of carbon dioxide believed to exist naturally in the environment may be seen in the previous graphic. The proportion of anthropogenic CO_2 needs to be weighed into consideration relative to naturally occurring reservoirs. We should take note that the general mood created in the media and elsewhere relies on what may seem to be a tiny proportion of the total CO_2 residing in the atmosphere at any time – that represent-ing human activities' carbon additions. According to this chart, we should consider the following:

- Earth's landmasses absorb an average of 1.5 GT of excess CO_2 per year.
- The atmosphere absorbs approximately 3 GT excess CO_2 per year.
- Anthropogenic CO_2 contributions total 6.5 GT per year.
- Total CO_2 exchange each year is 222 GT.
- Anthropogenic CO_2 represents $6.5 - 222 \times 100 = 2.9\%$.
- Actual net CO_2: $6.5 - 1.5 - 3 = 2$ GT.
- Net annual total percent of added anthropogenic atmospheric $CO_2 = 0.93\%$.

Despite the common impression by the populace that human activity is responsible for most atmospheric carbon dioxide, in fact, by our estimate above, only 2.9% of the total global CO_2 exchange (see later in this chapter) is attributable to the combus-tion of fossil fuels. In other words, it aligns fairly closely with the findings of Canadell et al. Others shade this figure upwards a little; for example David J. C. MacKay puts it at 3.27% [16], while some place it beyond this, but generally we don't find estimates of fossil fuel CO_2 to far exceed 5% of the total amount exchanged.

However, because it is this portion that is being added on an ongoing basis to the total that occurs naturally, it is this, therefore, that is a source of concern. Thus, in essence, the basis for the continuing disagreements about people's influence on climate can be further traced to this one statistic alone, and by what degree, if any, changes at this level have on the climate.

Whether the annual addition of this much CO_2 into the environment can be curtailed to a realistic degree cuts to the heart of the equation, if this should be required. Refer again to Fig. 2.8. It is instructive to pay careful attention to those sectors of the total that produce most of the CO_2 and aerosols; are we paying too much attention to some while largely ignoring others?

For example, if every form of fossil fuel transportation – land, air and sea – were to disappear tomorrow, only 13.5% of all major anthropogenic emissions would be eliminated! In a media sometimes out of touch with reality, you will not hear such a statistic; but you might hear that the automobile is the primary culprit for causing global warming.

Definitive positions to all such questions, so typical of almost everything in climate science, is more likely to encounter multiple answers than one single illuminating answer; they may depend upon whose argument strikes the individual as the most persuasive. Perhaps a more important question is whether we can all finally agree upon whether there are potential hazards awaiting us, and if there is anything we can do to stop them.

Hydrogen Fuel Cells and Electric Cars

"Hydrogen is high in energy, yet an engine that burns pure hydrogen produces almost no pollution."

– renewableenergyworld.com

"The electric car finally seems to be on the verge of breaking through, offering significant environmental benefits, especially in urban areas. Innovative business models are on the way which should boost consumer acceptance and overcome the remaining barriers, such as high battery costs, green electricity supply and charging infrastructure."

– European environment Agency

"Much touted as the solution to the automotive byproduct of CO_2, these currently fashionable technologies may not be all that they might seem, since each still requires the generation of electricity. In the case of hydrogen, this gas is generated by electrolysis and thus requires a significant electrical power supply; rechargeable batteries also require charging from the power grid."

"Similarly fashionable solar and wind power technologies are not yet realistic as major suppliers of electricity; as significant sources they are severely limited. Hydroelectric power is, however, a significant source of electricity, representing about 92% of the total green energy production, and about 24% of the world's total electricity, but only 8% of the total in the US."

– (Source: waterencyclopedia.com)

Nuclear electricity generation is still minimal in most countries. Thus the production of electricity mostly requires the burning of fossil fuels. Although at present loads it is possible to take advantage of off-peak hours when electricity might otherwise be wasted, for any large-scale automotive reliance on electric power, the problem brings us right back to where we started. Thus, the dream of entire highways running on these alternate technologies is still far away. Perhaps it never will be a reality.

Heat Sinks

The oceans and landmasses, in addition to being able to store large quantities of carbon dioxide, also have the capacity to act as vast "heat sinks," and much of the excess warmth from the Sun and greenhouse gases can be effectively buried. However, there are consequences from such storage. Because the ocean has far greater heat storage capability than the land, it has some of the most profound influences on much of what happens in Earth's climate. What is buried in the oceans drives the deep ocean currents, with significant consequences for the climate, all of which we will examine in due course.

Carbon and heat sinks are often linked. One study by S. Sitch et al. [17] proposed that carbon dioxide absorption by Earth's natural carbon sink will be compromised by increased temperatures beyond that which Earth can absorb, leading to greater influences of anthropogenic ozone due to the damage it causes to plant photosynthesis. The EPA (Environmental Protection Agency) has published an extraordinarily detailed analysis of the absorptive capacities of greenhouse emissions by Earth's natural sinks [18].

However, other researchers have come to the opposite conclusion. William Happer, in his paper on greenhouse gases [19], explained that the amount of atmospheric CO_2 was at the low end of what would be tolerable for life on Earth by a factor of at least 14 times the present levels. He also argued that the outgassing of CO_2 from the oceans had always *followed* temperature increases in the distant past, rather than the other way around. Thus, regardless of the formulae he used, this is a simple acknowledgement that Earth's heat sinks cannot simply "bury" all excess heat, any more than carbon sinks can soak up all excess concentrations either.

Happer's main argument was that variations of CO_2 concentrations were a natural reaction to temperature increases and decreases. During historical times, CO_2 concentrations were far greater than today, and, Happer argued, life flourished. Happer also seemed to concur with Nelson's findings on CO_2 warming effects, allowing only a 1°C increase with a doubling of present CO_2 levels (as have other scientists, including Richard Lindzen; see Chap. 6), although his proposed theory of reactive release of CO_2 seems incompatible with what we know.

Nelson had also mentioned the potential of Earth to absorb much of the increased heat in a process termed "masking." However, if he did not seem to take possible unseen detrimental long-term side effects of such processes into consideration, that may not be the correct assessment of his position. Also he did not discuss how much additional CO_2 might be released by the resulting increased ocean temperatures. Nevertheless, perhaps we would need to imagine a considerable absorption, over perhaps hundreds or thousands of years, before the accumulation might appear significant in comparison to the immense potential reserve of the oceans. The issue remains, in the IPCC's own term, "undocumented," and thus highly controversial at best. However, we must first consider in the next chapter two areas of study, clearly related though often confused with one another. Ultimately they are very different species.

References

1. Water vapor: feedback or forcing? Real Climate. Online: http://www. realclimate.org/index.php/archives/2005/04/water-vapour-feedback-or-forcing/
2. Graphic: effective temperature and comparison of the Sun to a black body. 2006. http://en.wikipedia.org/wiki/File:EffectiveTemperature_300dpi_e.png
3. NASA (1971) Earth albedo and emitted radiation. ntrs.NASA.gov/archive/nasa/casi.ntrs.../19710023628_1971023628.pdf
4. Andrews T, Forster PM (2008) CO_2 forcing induces semi-direct effects with consequences for climate feedback interpretations. School of Earth and Environment, University of Leeds, UK
5. Data Source: Keeling CD, Whorf TP (1998) Atmospheric CO_2 concentrations (ppmv) derived from in situ air samples collected at Mauna Loa Observatory. Scripps Institute of Oceanography, Hawaii; Neftel A et al (1994) Historical CO_2 record from the Siple Station Ice Core, Physics Institute, University of Bern, Switzerland
6. Global temperatures, from resource of graphs based of numerous satellite measurements. Climate 4 You. 2010. http://climate4you.com/GlobalTemperatures.html, http://climate4you.com/GlobalTemperatures.htm#Outgounglongwave radiation global
7. Nelson TJ (2011) Cold facts on global warming. Sci Notes. Online: http://www.randombio.com/co2.html
8. Samanta A, Anderson B, Ganguly S, Knyazikhin Y, Nemani RR, Myneni RB (2010) Physical climate response to a reduction of anthropogenic climate forcing. Earth Interact 14(7):1–11
9. Lacis AA, Schmidt GA, Rind D, Ruedy RA (2010) Atmospheric CO_2: principal control knob governing Earth's temperature. Sci Mag. http://www.sciencemag.org/content/330/6002/356.abstract
10. Watkins T. Global warming on Venus. Department of Economics, San Jose State University. Webpage: http://www.applet-magic.com/GWvenus.htm

11. Glassman JA (2009) The acquittal of carbon dioxide. Online: http://www.rocketscientistsjournal.com/2006/10/co2_acquittal.html
12. Mackay R (2007) Rhodes Fairbridge, the solar system and climate. J Coast Res 50
13. Robinson AB, Robinson NE, Soon W (2007) Environmental effects of increased atmospheric carbon dioxide. J Am Phys Surg
14. Canadell JG, Le Quere C, Raupach MR, Field CB, Buitenhuis ET, Ciais P, Conway TJ, Gillett NP, Houghton RA, Marland G (2007) Contributions to accelerating atmospheric CO_2 growth from economic activity, carbon intensity, and efficiency of natural sinks. Proc Natl Acad Sci USA 104(47):18866–18870. http://www.pnas.org/content/104/47/18866.abstract
15. Global flows of carbon. NASA graphic. Online: http://science.nasa.gov/earth-science/oceanography/ocean-earth-system/ocean-carbon-cycle/
16. Source: Prof. MacKay DJC, FRS, University of Cambridge, and chief scientific adviser to the UK Department of Energy and Climate Change
17. Sitch S, Cox PM, Collins WJ, Huntingford C (2007) Indirect radiative forcing of climate change through ozone effects on the land carbon-sink. Nature 448(7155):791–794
18. 2011 U.S. Greenhouse gas inventory report. Environmental Protection Agency. epa.gov/climatechange/emissions/usinventoryreport.html
19. Happer W (2011) The truth about greenhouse gases. Global Warming Policy Forum, briefing paper #3, 17 Aug 2011

3. Climate and Weather

Because many people confuse climate and weather, this has resulted in misconceptions about cause and effect between two phenomena that are actually pretty much unrelated. It is possible to have local cooling during times of global warming, and vice versa. Although the two are indeed related on some levels, ultimately, long-term climate patterns are established by entirely different factors than what we experience as short-term regional weather.

Earth's Energy Budget

The energy from the Sun that warms planet Earth is not distributed equally across the globe; regional temperature is determined primarily by latitude. Towards the poles, where the planet presents its most oblique face, the same amount of sunlight is spread out over relatively much wider areas than it is at the equator. Here, the face-on incidence of the incoming solar energy is closest to maximum concentration per square meter. As a consequence, the warming of the land and oceans is far greater here than at the poles, where at times no sunlight falls at all.

Through the angle of incidence, absorbed solar radiation is approximately 330 W/m^2 at the equator versus 150 W/m^2 at the poles. However, the rate of heat *lost* back to space is consistent from all points over the globe, creating zones. This also affects the circulation of seawater, as colder water displaces warmer.

The laws of physics dictate that a balance of incoming versus outgoing energy must be maintained. What has become termed 'Earth's energy budget' sums up that process. It is impossible for one part of the planet not to share the consequences of energy absorption or loss by another. In this respect, ultimately what is one of the great drivers of climate – arguably the greatest

A. Cooke, *Astronomy and the Climate Crisis*,
Astronomers' Universe, DOI 10.1007/978-1-4614-4608-8_3,
© Springer Science+Business Media New York 2012

byproduct of the warming effects of the Sun – comes about first in the atmosphere and eventually from energy stored deep within the sea.

Weather and Climate

Part of the problem in understanding climate relates to the regional nature of weather patterns *within* the larger realm of climate. The public typically confuses the two. ("It's been a really cold winter here in Kentucky – so much for global warming ...."). It is quite possible to experience extreme conditions for the short term when the overall global trend is exactly the opposite. And global climate conditions can vary substantially between the hemispheres, let alone the continents.

The atmospheric component of that driver comes about in two different ways, although the resulting winds will be felt in the same way and are both the result of air pressure differences. The simpler of the two to understand is what takes place when uneven localized solar heating over landmasses exposed to the greatest amount of sunshine initially creates pressure differences within the air above. Heated air being less dense than cold air has a correspondingly lower pressure. Subsequent wind currents are the result of the 'Pressure Gradient Force' (PGF), where zones of high pressure rush to normalize pressure in those of correspondingly lower pressure.

Because clouds reflect some energy back into space, cloudy regions will obviously be less warmed than those in full sunlight (such as deserts and other arid landscapes), and thus less likely to produce wind. Anyone who lives in a more temperate region close to deserts is familiar with the power of this phenomenon; the author knows well those hot early fall and cold winter desert 'Santa Ana Winds' – always moving westwards off-shore from the desert, in contrast to normal eastward on-shore conditions, keeping frequent low 'marine layer' (low clouds and fog) out at sea. Obviously, the most pronounced pressure differences over the smallest areas directly affect wind speed (Fig. 3.1).

The second and more complex process forms other larger low-pressure systems. In the hot equatorial regions wide air masses surrounding the globe are swept upwards, in a region termed the

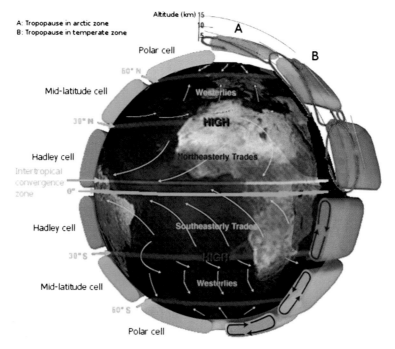

A: Tropopause in arctic zone
B: Tropopause in temperate zone

Altitude (km) 15

Polar cell

Mid-latitude cell

Hadley cell

Intertropical convergence zone

Hadley cell

Mid-latitude cell

Polar cell

60° N

30° N

0°

30° S

60° S

Westerlies

HIGH

Northeasterly Trades

Southeasterly Trades

HIGH

Westerlies

A

B

Graphic courtesy of NASA.

FIG. 3.1 Global atmospheric circulation (Graphic courtesy of NASA)

Intertropical Convergence Zone, or ITCZ. This behaves effectively like a giant evaporator, extracting moisture during the air's ascent. During the height of summer, the ITCZ can erupt into monsoon conditions around equatorial regions, as well as being the birthplace of hurricanes/cyclones. As the hot columns of air reach altitudes averaging 6 miles (~13 km), they begin to dissipate into ever-wider air masses that descend northwards and southwards in bands, most notably nearest the equator as 'Hadley cells,' but also as lesser mid-latitude and polar cells. Here, either they are heated again across southern deserts (spawning the easterly lower troposphere trade winds that return hot air back to equatorial regions), or cooled over northern deserts and tundra.

From the poles the air is cooled and heads back towards the equator in streams at low altitudes (see later in this chapter). The placement and altitudes of adjacent landmasses ultimately determine the patterns of motion and energy of these winds, as well as profoundly affecting their speed and direction. As the air twists and turns, it forms into low-pressure regions moving away from regions of high pressure. Familiar weather patterns emerge as low moist air masses spiral upwards, which may become laden with clouds and water droplets that often lead to rain; by contrast, high-pressure systems produce dry conditions.

It is easy to understand how different pressure gradients can be set up by the different responses of heat absorption on land and at sea. Landmasses respond by rapidly warming near the surface; the oceans absorb the heat at a significantly deeper level allowing the upper layers to remain cooler than the land. High and low pressure weather systems are the result. The upper water of the oceans is influenced by these air circulation patterns, especially in the formation of currents near the surface, as well as the forms that land dictates the open water take; in many ways, these ocean current mirror the winds above.

Coupled to Earth's rotation, major wind directions tend to move with that of the planet, rather than against it, through the process known as the Coriolis force. Interestingly, this force is at its maximum strength at the poles, and forms a counterbalance to the pressure gradient forces (PGF). This force is also connected to those pinwheel effects of cyclonic low pressure areas, the direction of rotation being determined by the hemisphere each occupies. Thus, weather systems take an easterly movement across most of the globe, corresponding to the rotation of Earth, and which we experience as approaching storm fronts or periods of stable fine weather. These ongoing evolving atmospheric conditions produce familiar day-to-day weather, the primary systems appearing as those large eddying pinwheel cloud formations we see in satellite images. Their spiral rotations take place because Earth's equatorial regions rotate faster than nearer the poles.

The Coriolis effect (after the nineteenth-century engineer Gustav-Gaspard Coriolis) might best be considered in the context of the Newtonian laws of motion regarding rotating bodies. In such a body, the rightward inertial force causes a deflection leftward relative the reference frame, and vice versa, and accordingly must be included in the total analysis of its motion.

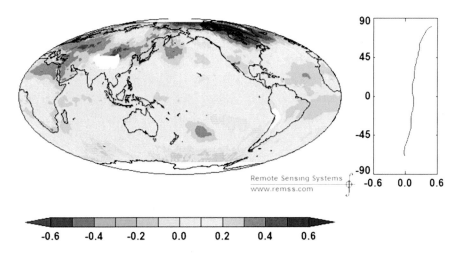

Fig. 3.2 Global temperature trend from 1979 to 2010: Sea level – 3,000 m (Image courtesy of Remote Sensing Systems. MSU data are produced by Remote Sensing Systems and sponsored by the NOAA Climate and Global Change Program. Data are available at www.remss.com)

Much has been made of the potential of the changing climate to increase the number and severity of storms, particularly hurricanes and cyclones. As with almost everything in climate science, disagreements are the norm. Various studies [1] have various conclusions. These range from positive, neutral and negative connections to warming temperatures, which further indicates the fractured discourse. On the surface of it, logically it would seem that since measured temperature increases have been greatest at increasingly northern latitudes, and less so toward the tropics, and even slightly *lower* towards Antarctica (Fig. 3.2), the reduced extremes between the two should result in smaller atmospheric pressure differences and thus *fewer* storms! But as is so often the case in this field of research, theory and actuality do not necessarily coincide.

One of the most significant fields of climate study, still far from settled, relates to clouds. The formation, types and quantity of clouds (along with rain and snow) have profound effects on atmospheric temperature zones, together with the resulting pressures and motions. Being endlessly variable, determining their precise effect is a moving target, something that has hampered definitive conclusions on their relative contributions to climate.

They act as a reflective blanket and are considered to contribute a negative feedback overall, despite the opposite blanketing effect below.

On average, covering approximately 50% of Earth's surface at any one time, it is easy to see how small variations in the total amount or placement of clouds could impact the weather or even the climate and remain one of the most undetermined aspects of the entire subject. (We will return to the subject of clouds, especially in regard to the study of theorized influences of cosmic rays upon their formation in Chap. 12).

Meanwhile, the movement of deep lying ocean movements is a separate matter again, working in many ways independently of the visible weather conditions we see in the atmosphere and at the ocean surfaces. However, certain attributes of each are interconnected and work in tandem, causing some dramatic larger periodic weather conditions.

Ocean Currents, Temperatures, and Oscillations

Ocean temperature and density are also tied up intimately with deep ocean currents. Density is primarily dependent on the quantity of salt dissolved in it, becoming less so as it releases salt compounds to form ice sheets and icebergs. As the increasingly dense saltwater descends in depth it is replaced by less salty water above. This, in turn, by the same repetitive process, initiates a deepwater current motion that forces deep lying cold water from the poles towards the equator, and most especially from the north.

Density is further connected with temperature, since water becomes less dense as it warms. Following the natural channel-like formation leading to the Atlantic Ocean, the cold salty water begins its journey southward in a circulation pattern termed thermohaline circulation. It is because of the formations of the continental landmasses surrounding the North Pole, and the consequent feeding of freshwater from the accumulations of snow and ice, that this deepwater motion is driven from the north instead of the south. The entire system, which will ultimately work its way around the globe, is known as the great global conveyor. Over a period of about 1,000 years, much of Earth's deep seawaters will have completed just one circulation:

- As it passes continental landmasses, the cold water is partially warmed (and loses some density), reaching past the equator and far into the Southern Hemisphere, eventually passing Antarctica to be cooled once again.
- Traveling north again, it splits and flows into two branches on each side of Australia – one east of Africa and the other up towards Japan via the Pacific Ocean.
- Once again, as the water flows great distances from these frigid regions, it gains warmth, and in doing so loses salt, which causes it to rise slowly nearer to the surface.
- In what may appear to be a kind of aquatic somersault, this warmer water doubles back on its tracks and rejoins into one stream traveling past the Americas and the continent of Africa on its way back to the North Pole, where it will start the journey all over again (Fig. 3.3).

Beyond this sequence of events, some significant variations can occur that are dependent on the degree that warm or cold waters predominate. Specifically, the warm and cold waters do not gradually mix with depth but rather co-exist in semi-separate states, the relatively shallow division between them known as the thermocline. Although the behavior of these is distinctly separate, variations in the placement and slope of the thermocline can produce large-scale oscillations in the oceans, the most significant being the El Nino Southern Oscillation (ENSO).

El Nino events (the product of ENSO conditions) are accompanied by low atmospheric pressure in the Pacific (with correspondingly milder trade winds) and high pressure in the Indian Ocean (coincidentally, corresponding to the double Great Ocean Conveyor streams after it splits into two parts). ENSO events usually trigger heavy rains from the eastern Pacific (western Americas) and drought on the western side in places such as Australia. Typically, ENSO events are cyclical in nature, occurring on average every 5–7 years.

To understand the fundamentals of how ENSO events come about, the diagrams (Fig. 3.4a–d), should make it more clear; it may be seen that atmospheric pressure has a direct correlation with the

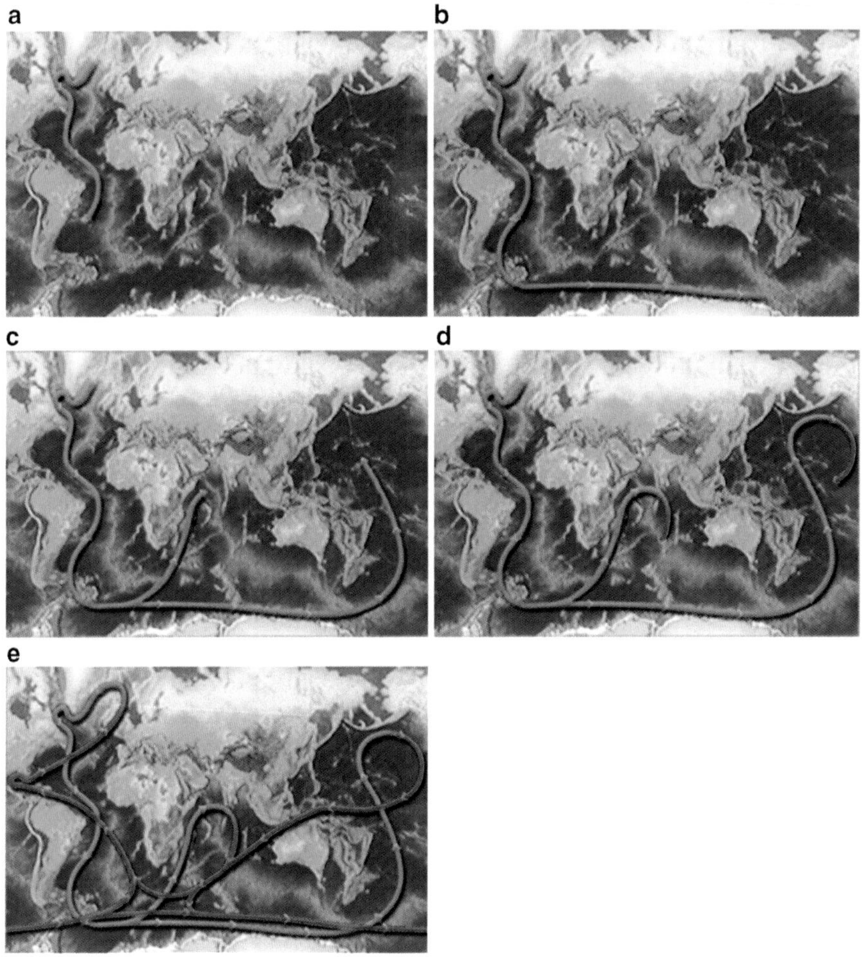

FIG. 3.3 (a–e) The great global conveyor (Graphic images courtesy of NOAA)

position of the thermocline. When it is low, warm water, normally kept west of the Americas, moves to the east much closer, bringing warmer winter conditions and more storms to the western American coasts (especially in South America), along with drier, warmer conditions to the mid-North American continent. Controlling factors of the cycle are not yet positively determined, although it has been linked to the 11-year solar cycle in a 2011 study appearing in *Science* [2] (also see Chap. 4 and 7).

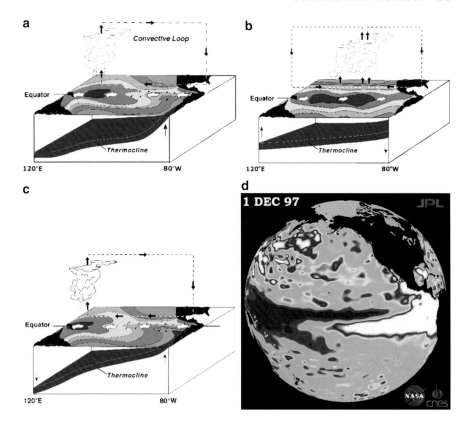

FIG. 3.4 (a–d) El Nino conditions. (a) Normal situation; winds keep warm water toward west Pacific. (b) El Nino developing; warm water moves east La Nina conditions: Warm waters further west. (d) Classic pattern of large El Nino event (Images (a–c) courtesy of NASA/PMEL/TAO; (d) NASA)

It is not known if or how present changes in the climate might (or might not) be linked to ENSO events. This is still being investigated; researchers have mixed positions at this time. Regardless, their cyclic nature appears independent to climate overall, with many factors driving it. Kevin Trenberth, however, of the National Center for Atmospheric Research, believes there is a significant connection between the two, theorizing that El Nino events are the natural mechanics of Earth's tropical regions eliminating excess heat. However, many uncertainties and questions remain, and the link has not yet been isolated and confirmed.

The opposite of El Nino conditions are termed 'La Nina' events, where the thermocline becomes steeply angled and warm waters are thus pushed westwards; the mechanics of these are plain to see in Fig. 3.4c. Under these circumstances, the situation is basically the reverse of El Nino conditions, whereby the Americas experience colder, drier conditions, and many locations on the opposite side of the Pacific experience warmer, wetter seasons.

It has been reasoned that the ongoing melting of glaciers in higher northern latitudes will result in a change to the salinity of the upper seawater relative to that forming the deep waters of the 'conveyor.' In turn this would effectively blanket it and push warmer waters southward. It is argued this would bring unforeseen consequences, including radically colder temperatures in Europe, as well as changes to El Nino and La Nina events, because these would be more likely to result in climatic changes, as opposed to shorter-term variety of weather. Exact scenarios remain unknown, as are the multitudes of possibly altered factors and their combinations. However, these do illustrate the degree to which the oceans affect Earth as a habitat for life.

Nearly as significant is the North Atlantic Oscillation (NAO), which is related to the Arctic Oscillation. Unlike ENSO events, those of the NAO occur entirely from complex pressure differences within the atmosphere, and thus, being primarily wind driven do not involve changes to the thermocline. Although it has no periodicity, if it should occur simultaneously with an El Nino event, the combination can produce extremely cold winters in Europe, accompanied by extreme weather conditions (Fig. 3.5).

Other Unseen Solar Connections to Earth's Biosphere?

Some researchers have recently proposed a very different vision of climate change, based on indirect solar amplification of thermocline circulation, with the possibility that relatively small direct changes in solar irradiation acting as a catalyst to changes in the circulation may be responsible. This was according to a 2010 study (Swingedouw et al.), that used new techniques of analysis of historical solar and climate patterns in relation to the North Atlantic Oscillation [3]. If this proposal seems a radical step, it is not the first time such a link has been suggested, especially since slight

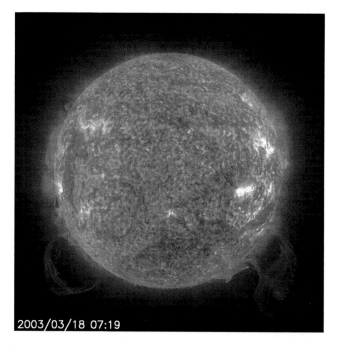

2003/03/18 07:19

Fɪɢ. 3.5 The Sun with huge prominences (Image courtesy NASA/SOHO)

differences in solar irradiation from year to year have long been discounted by the IPCC in affecting climate significantly. In this new theory, it was suggested that changes to the wind patterns are directly linked to small changes in solar forcing, and ultimately to climate change itself.

However, such connections remain speculative, undetermined and not yet proven, although it is conceivable that they exist. Certainly, a proven link between ENSO and NAO oscillations to minor solar variations would cause a major shift in climate science. (Also refer again to Ref. [2].)

Additional resources on all aspects of this and related matters are plentiful. In one example [4], climate was linked to the Pacific and Atlantic ITCZ climate variability, including rainfall in the semi-arid northern parts of Brazil. In another example [5], extended possibilities along these lines is a rather startling summation that the warming of the later twentieth century was caused by a response to warming of the oceans and *not* directly to increased atmospheric anthropogenic gases over land masses. Although the authors of this study do allow for greenhouse gases to be partly responsible, they believe that increases and warming of atmospheric water vapor has led to greater penetration and subsequent absorption of long-wave radiation into the oceans.

Interactions of the Atmosphere and Sea

Earth's atmosphere is considered to be a 'chaotic' system, which is why wild swings in weather patterns of small pockets and eddies do not conform to any sequence that can be computed for more than a few days ahead. Climate, which is a longer-term phenomenon, is in many ways, at least to some degree, more predictable and 'linear.'

Because the NAO is considered a wind-driven oscillation, we should examine how large-scale zones of pressure deviate from even distribution, ultimately determining long-term weather patterns. However, ENSO events are also connected to the atmosphere through pressure interactions with the thermocline. And we already know that at the lowest altitudes, the disruptive effects of landmasses, especially mountainous regions, cause air to form into eddies of various shapes, sizes and descriptions; these affect localized, relatively small-scale and short-lived weather conditions.

At higher elevations, however, the effects of the terrain on the atmosphere are far less pronounced, in contrast to the atmospheric conditions below. What takes place higher in the atmosphere significantly controls the longer-term weather periods we recognize, including entire seasons, such as hot summers, wet and cold winters, and so forth. These are features of so-called Rossby waves (Fig. 3.6), which constitute the defined regions of pressure in the atmosphere at these altitudes. Complex patterns of friction (termed 'shear') cause variations in inertia of upper air masses to form unpredictably from smaller weather systems below. These, in turn, cause significant irregularities of pressure, as in Fig. 3.6c.

Many factors affect the formation of Rossby waves, large irregularities in what would otherwise be evenly distributed pressure zones by latitude. On Earth, rotating air masses have a predominantly westerly motion, due to the Coriolis force, which, according to latitude, correspond with different inertia to form waves. The speed of these air masses is also partly driven by temperature (and thus, pressure) differences between the poles and equator.

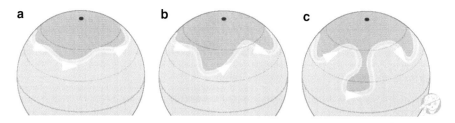

FIG. 3.6 The formation of Rossby waves, (a), minimal circulation irregularities, (b) increasing circulation regularities (c) Rossby wave formations (Courtesy of Wikipedia (de:User:W))

These speeds might be considerable, often reaching several hundreds of kilometers per hour - a swirling global upper wind current better known as the jet stream.

It has been theorized that increasing heat in the upper atmosphere causes the strength of the jet stream to increase. Though this may be to a relatively small degree, it protects northern latitudes from extremely cold arctic air masses that would otherwise descend southwards. As the jet stream weakens from decreased solar activity it fragments, thus allowing the frigid arctic air to flood into more southern regions. The resulting complex conditions are a good example of non-linear response, completely unpredictable in normal climate forecasting. If these effects could reliably be demonstrated, the scientific value would be profound. Could this possibly be, in fact, what could have occurred repeatedly over many decades to contribute to the period known as the Little Ice Age, from approximately 1550–1850? (see Chap. 5).

Although the path of the jet stream tends to fluctuate according to the Rossby wave formations, it corresponds with variations in the displacement of the thermocline, which ultimately drives El Nino and La Nina events. The position of the jet stream is also affected by the transport of heat within the oceans and very high mountainous regions. It is also here that large loops of spinning air (as in Fig. 3.6c) separate from the whole and move towards the equator to create larger weather patterns, including hurricanes and tropical cyclones. Differences in seasonal weather at locations of similar latitudes seem to be related to these waves and their interaction with warm waters [6].

We can observe similar, although not quite parallel, conditions on other planets, notably Jupiter (Fig. 3.7), where its parallel

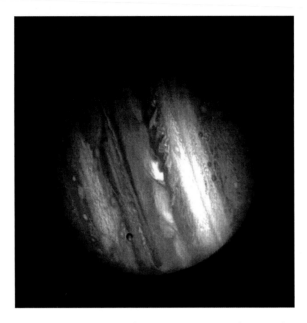

FIG. 3.7 Jupiter (*Voyager 1*, 1979) (Image courtesy of NASA)

stripes, diagonal streaks and rotating spots – actually Jovian storms (the most famous being the Great Red Spot) echo what is occurring in our own atmosphere, at least superficially. However, in the absence of Earthlike protruding terrestrial landmasses, similarities to the wide variations of Rossby waves in atmospheric zones of pressure are far less pronounced.

Tracking Sea Surface Temperatures

Rising sea surface temperatures (SST) represent another consequence of global warming, as might be expected. A casual inspection of available reference materials can be misleading, as it is common to include average land temperatures with those of the oceans. However, an accurate assessment of ocean surface temperatures alone is important for our full understanding, since both have entirely different base temperatures and mechanisms of warming and cooling. Water, a moving mass, is also subject to changes in volume relative to temperature.

Unfortunately, graphs readily available from NASA and NOAA (the National Oceanic and Atmospheric Administration)

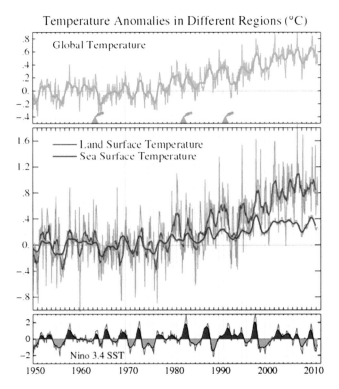

FIG. 3.8 Global temperature anomalies (Graphics courtesy of Columbia University, from data provided by NOAA)

do not show separate *long-term* ocean temperature changes. The custom is to provide month-to-month temperatures (with commentary), a practice that allows many claims on independent climate sites, dubious or otherwise. Fortunately, graphs and graphics for every conceivable point of curiosity may be found online at NOAA's affiliate organization at Columbia University [7].

Careful analysis of SST's quickly reveals that their average readings fluctuate much less than those of land. From Fig. 3.8, we can also see that there has not been much more than a 0.2°C increase since the early 1950s. From this graph, we can also see that SST's have appeared to decline overall from their high point around the time of the El Nino of 1997/1998, up until around 2008. Certainly there was a significant upward spike in 2010, as this graph shows. Whether or not it is an anomaly or a trend remains to be seen, since many dips and peaks can accompany overall temperature increases just as easily as they do declines.

We should also consider the thermal expansion of water, since this is often cited as a leading cause of sea level rise. When water freezes, its density decreases just before it does so (916.8 kg/m³ vs. 1,000.0 kg/m³), so it expands to take up more space than in its liquid form. On melting, the density returns to maximum (1.00 m³/kg), at 4°C. However, water density then decreases again with increasing temperatures, meaning its volume again increases, although in order to register an increase in volume of just 1% (1.01 m³/kg), water temperatures would need to heat to an average of 35°C – bathwater temperatures!

SST readings currently average 16.1°C and are considered to be presently 0.47°C above 'normal.' Thus, it would take a rise of almost 20°C to increase total volume by 1%. Of course, this does not take into account any differences between seawater and fresh water, or the ways in which the oceans are spread throughout the total area around the globe, those large shallow depths surrounding coastal regions, or the prior displacement of water by the ice. But one could certainly be forgiven for wondering if an increase in temperature of a fraction of a degree would cause significantly measurable differences in sea level. However, with the average depth of all of the oceans being 14,000 feet (4,267 m), according to NOAA, this translates, superficially at least, to a rise in sea level of some 5 feet. We can calculate this once we know that present average SST is 16.1°C, its specific density is 999.2 kg/m³, and at 1°C higher it is 998.3 kg/m³, by the simple formula:

992.2 – 998.3 = 998.84 (specific density decrease of 0.36).

Thus:

$$999.2 \times 14,000 = 14,005 \text{ feet}$$
$$998.84$$

What we are looking at is a rise large enough to cause major harm to coastal communities. This is just a basic reference point for estimating sea level rise based on a constant ocean depth of 14,000 feet. Therefore, based on any of the contributing factors, the actual rise could be less, or even more, depending on any number of unknowns. Luckily, there are more sophisticated means to look into this.

The JASON Satellite Radar Program

Future sea level projections have been constantly revised downwards as new information emerges. In fact, luckily it happens that the hypothetical amount of 5 feet resulting from a 1°C increase actually is hugely more than has been estimated in recent NASA projections. These have been stated as a recent maximum of 3.27 mm per year, over a period representing increases in temperature no greater than 1°C (1900–2000), including a controversial and arbitrarily added 0.3 mm per year. That was an entirely new topic for controversy.

When it was reported that NASA and researchers of the U.S. Sea Level Group at Colorado University had agreed to the addition of that 0.3°C arbitrary adjustment to sea level readings [8], critics immediately seized upon the directive, accusing NASA of perpetuating a 'trick.' It was said that it would enable government funds to continue to be channeled into a failing climate change theory, specifically, that of human-induced warming. However, sea level researchers claimed it was necessary to compensate for the vertical upward shift of continental landmasses. Critics argued that this had been occurring for thousands of years (and referenced earlier), but never counted before.

Curiously, also, the measured rise in sea level increases appeared suddenly almost to double after initiation of the JASON satellite program in 1993. Measured calculations showed a change from an average of 1.8 mm per year (measured ground data up until year 2000), to the new JASON readings, averaging 3.27 mm per year from 1993 to the present. If we subtract 0.3 mm per year, the annual additional increase in sea level still turns out to be less than 3 mm, so it all depends on what amount we consider to be problematic. More importantly, it begs the question of what has been human induced, as well as what the addition means. The recent uptrend has been attributed to more rapid melting of glacial and polar ice, as well as expanding water masses due to increasing temperatures. However, recent satellite measurements indicate there may have been no increases in sea level since 2006; climate change controversy seems to be the gift that keeps on giving.

Excerpted from NASA, August 2011

"Like mercury in a thermometer, ocean waters expand as they warm. This, along with melting glaciers and ice sheets in Greenland and Antarctica, drives sea levels higher over the long term."

"... while the rise of the global ocean has been remarkably steady for most of this time, every once in a while, sea level rise hits a speed bump. This past year, it's been more like a pothole: between last summer and this one, global sea level actually fell by about a quarter of an inch, or half a centimeter."

"What does it mean? ... while 2010 began with a sizable El Niño, by year's end, it was replaced by one of the strongest La Niñas in recent memory. This sudden shift in the Pacific changed rainfall patterns all across the globe, bringing massive floods to places like Australia and the Amazon basin, and drought to the southern United States."

"... each year, huge amounts of water are evaporated from the ocean. While most of it falls right back into the ocean as rain, some of it falls over land."

Unsurprisingly (as we will continue to see is the norm in climate studies), other researchers can be found who contradict most, if not all, of the above! A group of European researchers was directly critical of the JASON conclusions (Ablain et al.), and challenged these findings [9]. Claiming the rise had been overestimated by as much as 60%, the study stated that projections should be revised downwards by 2 mm per year, not an inconsiderable amount. As such, it would change everything if confirmed.

As always, mixed statistics such as these provide more for continuing the ongoing debate than for clear resolution, although providing a rather different perspective of what really ought to be regarded only as an illusion of long-term stability. When it comes to the world's oceans and their influence on recent changes in the climate, it seems we may have previously taken them for granted. Couple this with the fact that the oceans account for 71% of the total area of the globe, and it emphasizes that we need to be careful in processing all temperature statistics in the same breath.

In regard to warming trends, the oceans do not mirror the land. Rhodes Fairbridge [10], as early as 1958, discussed three principle factors controlling sea levels: basin shape, volume of water in them, and variations in the adjacent landmasses. Apparently he did not deal with the effects of temperature, but was aware of a gradual overall sea level rise during the Holocene, occurring in repeated cycles of surprisingly short intervals of as little as a few decades, as evidenced by rapid changes in levels shown in tube-worm deposits on rock formations. The huge amounts he believed could be supported by evidence would be enough to swamp any low-lying regions, so we should be grateful that his studies showed decreasing regularity and level! However, according to his studies, we should always consider sea levels prone to change drastically at any time.

One last interesting possibility was raised (long ago), in 1974 by Nigel Calder [11], who proposed that the northern polar ice cap is constantly being fed by the Atlantic Ocean. The increasing size of reflective ice sheet causes more solar energy to return to space, which in turn freezes that part of the Atlantic that supported the polar cap growth. Because there is less liquid water to evaporate and become snow, the cap will slowly shrink, and the waters warm. Ultimately that same warmth produces a thaw in the Atlantic source of water for the polar cap, and thus its growth can resume, reversing the warming trend! Although an interesting, if far-fetched, proposition, it still surfaces today now and then. Robert H. Essenhigh, writing in the May 2001 Journal of the American Chemical Society, included this scenario quite firmly in his concluding remarks [12]. His credentials are impressive, although he is clearly a scientist who takes a 'skeptical' view of AGW. However, his opinion similarly does not appear to have been echoed in wider circles. This concept has generally faded from view, but it remains interesting, nonetheless.

Trying to Make Sense of Everything

It may be becoming increasingly clear that many of the factors of forcing and feedback involved in climate change are anything but predictably linear. Together, they interact and modify the characteristics of the collective group of factors, resulting in some becoming stronger, others weaker, or even canceling many altogether. Since numerous additional factors are 'chaotic' – in essence unpredictable in the mix – and still others follow linear patterns, it is not difficult to imagine how a change to any one of the linear or chaotic factors could trigger unforeseen consequences. It seems, in essence, that we depend on a fragile state of *near* balance.

To understand how the interactions of all contributing forcing and feedback factors are considered by scientists, let us, for any one of these factors, imagine a graph with high and low points connected by a curving line. This is termed a 'time series.' Figure 3.9 illustrates a time series of the North Atlantic Oscillation over 62-year period.

Now imagine similar graphs created for every other factor in the mix, such as precipitation, wind, temperature, etc. Smoothed graphs representing each can be studied for their separate characteristics. The key to all time series is the measurement of predetermined values for a specified period of time. However, as applied to climate research, a time series does not tell us much about the characteristics of energies within the signal itself. For this, we will need to extract such data by what is termed a 'Fourier Transform.' Fourier (1768–1830), a French mathematician and scientist, realized that it was possible to express virtually everything on any kind of timeline as an expression of interacting different harmonic amplitudes of sine and cosine waves. He realized that even saw tooth and square tooth waves could be broken down into a multiple of such waveforms.

A time series, such as in Fig. 3.9, can be seen as a series of points of amplitude at regular intervals – in this case determined on an overall annual cycle. The 62 different annual cycles can be broken down into a total of 31 harmonics, commencing at 62, then 31, 20.66, 15.5, 12.4, and so on, down to the final harmonic, which will be 2 by default, as in every time series. By applying Fourier's mathematical formula, the element of time is removed, and these harmonics can be represented as high and low points, revealing where the strongest inputs are located as factors of amplitude. Thus a new graph of amplitude versus frequency is produced, revealing the key characteristics of the whole, regardless of time.

It is usual to go further in extracting the primary information in a time series by forming a 'power spectrum.' By taking a Fourier

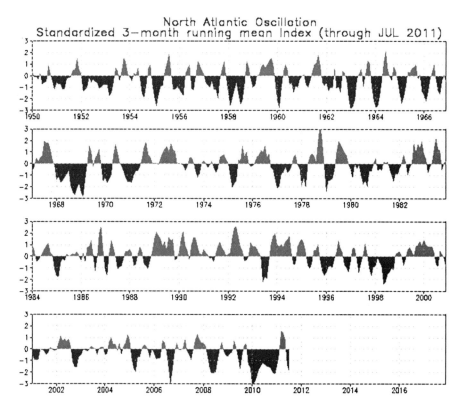

Fig. 3.9 NAO time series (Image courtesy of NOAA)

transform and squaring the amplitude of each harmonic, a power spectrum is produced. The amplitude of each harmonic frequency of a Fourier transform is squared to produce an exaggerated but clear outline of the strongest harmonic influences of the series. The individual squared harmonics may be shown as simple lines or columns, or the tops of each may be joined into a continuous series of spikes instead in a graph representation. Still other factors, random ('chaotic') parts of the whole, register as 'noise' of one kind or another (referenced as 'white' through 'red,' as in a color spectrum, red referring to low frequency factors), and as such, if not accounted for it can confuse the outcome and interpretations of these findings.

A good, comprehensive description by Melissa Ray Weimer about the underlying principles of Fourier transforms and power spectra may be found online (although a commercial page about

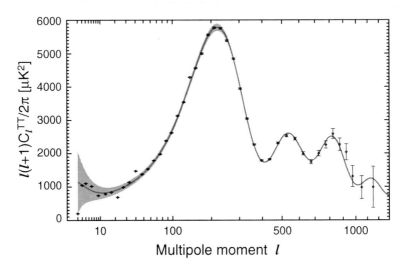

Fig. 3.10 Angular power spectrum of the fluctuations in the WMAP full-sky map – the relative brightness of the 'spots' versus the size of the spots (Graph courtesy of NASA/WMAP Science team)

software needs, written for DATAQ Instruments, Inc.) [13] for those interested in knowing more. For many readers, this article will be welcome indeed in relaying the essential functions in terms that are accessible, among the many that approach the subject as if every reader is a math major. In fitting conclusion to the chapter, Fig. 3.10 is a power spectrum of the cosmic microwave background discussed at the beginning of Chapter 2.

Power spectra represent a standard mechanism throughout science, as well as in numerous fields unrelated to science. For compressing and interpreting complex inputs into a coherent and compact format, they are of considerable value in climate studies. Here, so many factors exist to begin with, with many more variable in nature and others still not fully understood, and may be included as the research proceeds. Such mathematical coding remains related to climate modeling, a system of determining future directions of change. Should any factor change, potentially it is hoped to project its effect on the others. However, the problem is recognizing and understanding all of the factors involved, and this remains at the heart of the dispute of the accuracy of existing climate models. More about this in Chap. 8.

References

1. Henderson-Sellers A et al (1998) Tropical cyclones and global climate change: a post IPCC assessment. Bull Am Meteorol Soc 79(1):19; Sugi M et al (2002) Influence of the global warming on tropical cyclone climatology: an experiment with the JMA model. J Meteorol Soc Jpn 80:249–272; Emanuel KA (1999) The power of a hurricane: an example of reckless driving on the information superhighway. Weather 54(4):107

2. Meehl G, Arblaster J, Matthes K, Sassi F, van Loon H (2011) Amplifying the Pacific climate system response to a small 11 year solar cycle forcing. Science 325(5944):1114–1118

3. Swingedouw D, Laurent T, Christophe C, Aurore V, David S-M, Jerome S (European Geophysical Union, General Assembly) (2010) Natural forcing of climate during the last millennium: fingerprint of variability. Springer, New York

4. Kaspi Y, Schneider T (2011) Winter cold eastern continental boundaries induced by warm ocean waters. Nature 471(7340):621–624

5. Chiang JCH, Kushnir Y, Zebiak SE (2000) Interdecadal changes in Eastern Pacific ITCZ variability. Geophys Res Lett 27(22)

6. Compo GP, Sardeshmukh PD (2009) Oceanic influences on recent continental warming. Springer, New York

7. Columbia University ocean temperature graphs, 2012. http://columbia.edu/~mhs119/Temperature/T_moreFigs/

8. Colorado University Sea Level Research Group, statement, 2012. http://sealevel.colorado.edu/content/global-mean-sea-level-time-series-seasonal-signals-removed

9. Ablain M, Cazenave A, Valladeau G, Guinehut S (2009) A new assessment of the error budget of global mean seal level rate estimated by satellite altimetry over 1993–2008. European Geosciences Union

10. Calder N (1974) The weather machine. Viking Press, New York

11. Mackay R (2007) Rhodes Fairbridge, the Solar System and Climate. J Coast Res 50

12. Essenhigh RH (2001) Does CO_2 really drive global warming? Chem Innov (a Journal of the American Chemical Society) 31:44–45

13. Weimer MR. Waveform analysis using the Fourier transform. DATAQ Instruments, Inc. 1989. http://www.dataq.com/applicat/articles/an11.htm

4. The Variable Sun

The Sun is central to everything in our existence; it is also central to our potential demise, for many reasons. It is a relatively stable main sequence star, but proxy records from antiquity indicate that its output has varied often over past millennia (Fig. 4.1).

Early in the twentieth century, scientific inquiry had already begun into potential links of the Sun's activity to Earth's climate. This was at a time when climate change was far from the public's – even scientists' – consciousness, and intimate knowledge of the workings of the Solar System, let alone the greater universe, was still remarkably limited.

Of course, long before in prehistory, ancient man was well aware of regular periodic cycles experienced on Earth: the length of a day, the lunar month, tides, seasons, etc. Indeed, from their perspectives, they were central in a universe they believed they could comprehend. They did not know, of course that Earth circumnavigates the Sun, and that we are part of a Solar System of other planets orbiting the core of a galaxy made up of billions of other suns. Even more, that our Sun periodically compresses together into wavelike formations of other star populations to create galactic arms, or that periodically it moves through clouds of interstellar matter and even lesser densities of suns between those galactic arms. Primitive man also had no awareness, of course, that the Sun was 'merely' another star, just like the countless tiny points of light they looked up towards in the dark night sky, and certainly no awareness at all of the larger portions of an invisible electromagnetic spectrum: that part of it that they experienced as light and dark, warmth and cold.

We know now, of course, that we are part of a vast galactic system among a universe full of billions of other such galactic systems. It is only reasonable, therefore, that educated individuals might contemplate that some of the countless external astronomical forces might possibly be reflected in Earth's habitat and climate.

A. Cooke, *Astronomy and the Climate Crisis*,
Astronomers' Universe, DOI 10.1007/978-1-4614-4608-8_4,
© Springer Science+Business Media New York 2012

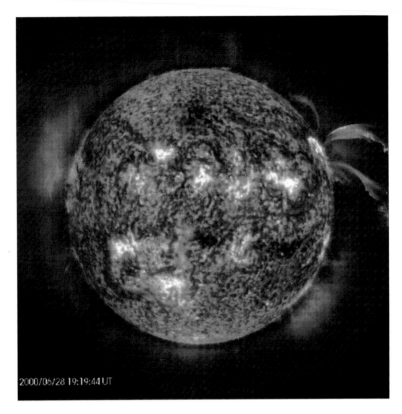

2000/06/28 19:19:44 UT

Fig. 4.1 The Sun with coronal mass ejection (Image courtesy of NASA. Although this image shows remarkable detail, the exposure has been reduced to a degree that it is obvious the color is not true to life)

As might be expected, controversies galore exist around all such types of research, much of which is routinely dismissed without even so much as fair consideration from some biased individuals. Naturally, some imitations of research masquerading as science are nothing but pure sorcery, unfortunately reflecting badly on other real science that could be legitimate. Although not all recent theories will stand the test of time, or the closer scrutiny that comes with that, we need to look into them if we are to sort the wheat from the chaff. At the very least, they provide thought-provoking concepts.

Without some knowledge of the behavior of the Sun itself, it would be hard to understand those aspects of it that have been implicated in recent studies on climate change. In 2007, Henrik

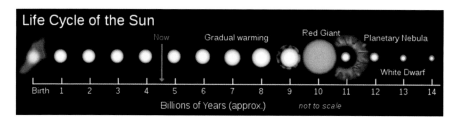

FIG. 4.2 Life cycle of the Sun (Graphic courtesy of Oliver Beatson)

Svensmark, director of the Center for Sun-Climate Research at the Danish National Space Center, commented that solar activity had been exceptionally high during the twentieth century, compared to the previous 400 years [1]. He also mentioned his joint discovery in 1996 of an apparent direct link with cosmic ray intensity and cloud formation, the latter having decreased by 2% at the height of each sunspot cycle [2]. Other researchers also have tried to show this link with cosmic rays and solar activity. (We will return to this in Chap. 12.) Svensmark has also expressed the opinion that climate models do "a poor job" in not including such possible factors in their input.

At about 25,000 light years from the Milky Way's center, the Sun is situated in the Orion-Cygnus Arm. Somewhat misnamed, this arm is really a spur, more akin to a isolated fragment between the main Perseus and Scutum-Centaurus arms. Historically, during the twentieth century, the Sun was considered to be more or less an average star, but has now been determined to be anything but ordinary. Apparently it outshines the majority of stars in the galaxy. Because of the abundance of some fairly exotic elements in the Solar System, it can be said with a fair degree of certainty that a supernova explosion must have occurred nearby, since only supernovae are capable of forming those elements. Presumably it was this explosion, close enough to a condensing cloud of interstellar matter, that triggered its final collapse. Thus began another cycle of star creation – whether singly or in a cluster – and this is likely how our Sun came into being.

Presently, about 4½ billion years later, the Sun finds itself a middle-aged star, on the cusp of beginning its long journey to red giant, senior status. We can see in the graphic (Fig. 4.2) that the

Sun does not have too many billions of years to go before its demise arrives, but suffice it to say, we have a little time left before having to worry about *this* particular form of climate change. When it happens, it will end life on Earth as we know it.

Although the surface (photosphere) temperature of the Sun is around 6,000°C, within its outer corona it is another matter entirely, with temperatures ranging from an astounding average 1.5 million degrees Centigrade to at least ten times that in places. The total energy emitted from the Sun has been calculated to be 3.8478×10^{-26} W, an almost inconceivable amount. However, constant variations in its output shade this figure slightly, but through recent millennia they are of the order of about 0.1%.

Therefore, if solar activity were to be implicated in recent warming, it would *not* be by direct warming effects alone, a pivotal factor in many scientists' views. In other ways it is this small amount that some researchers believe may hold at least some of the keys to climate variations. Stuart Clark, in a 2010 article, discussed the current dearth of sunspots, and its significance relative to the ways such activity might affect the climate [3]. Clark stressed that ultraviolet radiation in the Sun's spectrum may be at the heart of changes to Earth's climate and weather.

Sunspots

From quite early astronomical times, keen-eyed observers noticed from time to time that dark spots could be detected moving across the face of the Sun. Identified as true solar phenomena, they appear dark in comparison to the brilliant surfaces that surround them. As merely much *less* brilliant regions, their temperatures are still in excess of 3,000°C, even 4,000°C. Lying at the heart of research into solar variability, sunspots are major indicators of solar activity. Formed deep inside the Sun, they manifest themselves as the familiar dark blotches when they rupture on the surface.

With diameters of up to 40,000 km, their general appearance, as shown in Fig. 4.3, is quite typical, although this image captured a particularly prominent high point during a strong cycle. These irregular speckled blotches are each bordered by 'faculae,' elongated horizontal tube-like features. Grouped together, they form

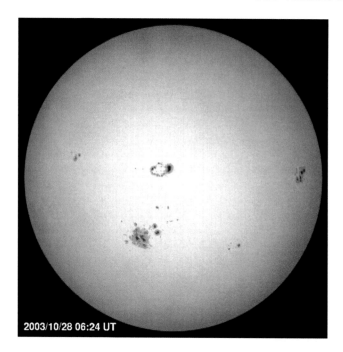

FIG. 4.3 Sunspots (Image courtesy of NASA)

the wide penumbrae around the dark spots themselves, their structure determined by the strong magnetic fields of the Sun. These surrounding regions are brighter than the sunspots but still less than the brilliant surrounding solar disc. Typically, the spots are short-lived features, most of them lasting much less than a week. They often 'hunt' in pairs, or even in multiples, typically gaining in size at the expense of their opposites.

When it was discovered that there are periods of greater, as well as minimal, sunspot activity, it was thought that because these regions must be cooler (indeed they are), this would signify slightly *reduced* solar output. However, the opposite is true, because of the increased irradiance of the surrounding faculae. Now known to directly affect the total solar output, faculae are more transparent to extremely intense radiation emerging from deep within the photosphere. Apparently this explains the increases in solar irradiance when many sunspots are present, when otherwise one would have expected the cooler inner regions of sunspots to reduce it. It turns out that radiation is greatest when

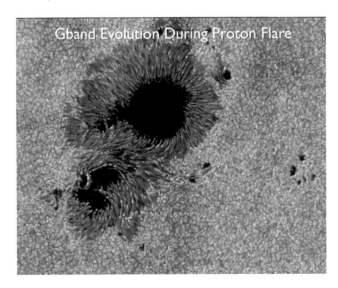

FIG. 4.4 Sunspot close up, 2006 (Image courtesy of NASA Hinode JAXA)

solar activity is at its height, and not necessarily when only *more* sunspots are visible.

In the next view, a close up of a pair of sunspots (Fig. 4.4), we can see not only the highly granulated surface of the Sun itself in stark relief but finally into a sunspot itself. Apparently, this surrounding granulation represents convective cells rising from the Sun's interior into the photosphere (the overall 'surface'). Faculae also may be clearly seen, their more linear structure obviously different from the surrounding granulated surface. The Sun's surprisingly rapid rotation of 27 days (similar to that of Earth's Moon but shorter by just a day) is sufficient to show the apparent paths of sunspots from day to day.

Unsurprisingly, considering the degree of contrast required in order to effectively register detail in solar imagery, beneath the 'surface' (photosphere) we cannot make out anything inside the spots themselves. However, there is another, far more profound reason they appear so dark. The Sun is composed primarily of molecular hydrogen; you may recall this is the primary ingredient of the interstellar medium from which it formed. We can only see on or near the surface of the Sun's photosphere due to the creation of negative hydrogen ions. Because at ever-increasing depths below the surface negative hydrogen ions are created in fewer quantities, light cannot escape much below it. Therefore sunspots, quite aside from being relatively dark compared to the even more brilliant photosphere that surrounds them, cannot reveal anything from deep within the Sun, regardless!

The temperature at the center of sunspots is about two thirds that of the photosphere (surrounding surface). The faculae certainly are still hot enough to be dazzlingly luminous if we did not need to reduce exposures radically in order to see the spots themselves. The surrounding solar photosphere would normally register as exceedingly brilliant white, despite the Sun's categorization as a yellow dwarf. This is a term that really neither describes its color as would be seen in space, nor its true status outside astrophysical protocol – not to be confused with brown dwarf or white dwarf stars that represent the last vestiges of most stars' life cycles and are truly diminutive in stellar terms. This is because yellow is its dominant wavelength in visible radiation, and dwarf refers to *all* such main sequence stars that have not yet become red giants in old age.

Superficially, at least, sea surface records would seem to reflect the sunspot record with the least complication, since the oceans do not have such compounding factors as urban areas, mountainous regions or desert regions, for example. Certainly, NOAA records from the mid-1800s until 1980 do seem to confirm a direct connection of sunspot activity to global sea surface temperatures (see Fig. 4.5). Regrettably, this graph does not take us to the present – the period since 1980 where the *real* controversy began. Regardless, a link to temperature and solar activity seems possible to draw within the time frame of the graph.

Solar Cycles, Minimums and Maximums

Because sunspot activity directly affects solar output, it readily can be argued that this link to climate underlies all other issues that divide those who maintain that modern climate change is human-induced (the position of the IPCC), from those who deny it just as staunchly, maintaining that it is primarily solar-induced – the believers and the non-believers. Indeed, certain recent measurements imply that we may have entered a cooling trend

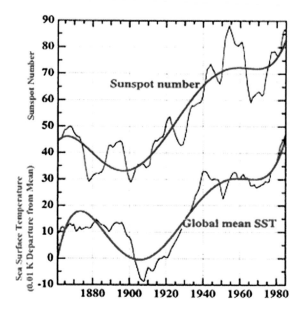

FIG. 4.5 Correlation between sunspots and sea surface temperature (Graph courtesy of NOAA)

due to reduced solar activity, but only more time will tell if that is so. Climate does not follow simple outlines, as there are usually spikes in all directions, regardless. One paper by David C. Archibald steps out on a 'solar' limb with the prediction that temperatures will actually *decline* 1.5°C by 2020 [4].

It has been determined that there is approximately an 11-year period separating one cycle of sunspot maximums to the next. These cycles usually overlap by a year or two. At maximum activity in these cycles, sunspot formation gradually migrates towards the solar equator. There have been 23 complete cycles since record keeping began, and we have now entered the 24th cycle.

The late Timo Niroma, an independent Finnish researcher, climatologist and student of the Sun, made a detailed study of the historic record. In the absence of any readily available peer-reviewed studies by him, it is difficult to assess his work, although he assembled some of the most detailed writings on sunspot cycles still available online on his personal website [5]. Going on record with much the same idea, he predicted that a new sunspot super minimum was on the way, which would result in a delay to the start of the new (present) cycle 24. He stated that activity would be extremely low, even to the extent that a new 'Maunder Minimum' (an historic time of greatly diminished solar activity) might be approaching. To date, the cycle does appear to be less active than normal, and appears it will be the smallest sunspot cycle in the past 100 years, according to the Marshall Space Flight Center (September 2011).

For those who wish to delve into the topic more deeply, Niroma's unabridged discussions about the workings of the Sun may be found on his website. Additionally, the website includes material that references peer-reviewed research. Only time will tell if any of his long-range predictions become reality, however, together with those of so many others who theorize low sunspot activity means lower temperatures. Niroma remains a somewhat mysterious figure, although he was respected enough to see his article, 'Solar behavior, and its influence on Earth's climate,' included in the Viewpoints and Technical Communications section in *Energy and Environment*, Vol. 20, 2009.

Additionally, in 2004, Habibullo I. Abdussamatov, head of space research at Pulkovo Observatory, investigated the 11-year cycle, as well as variations in solar radius, irradiance and activity correlations, concluding they are caused from deep within the Sun. Because of historic solar behavior patterns, he, too, concluded that the Sun has entered a cooling phase that will counteract any effects of anthropogenic warming [6] (see also Chap. 9, [7, 8]). However, Abdussamatov is no stranger to the contrarian view.

We are perhaps fortunate that we live in a time when soon we will see whether the hypothesis by some, that a marked cooler climate period is taking place through reduced solar activity during the next few decades, is correct. Thus, we can test the theories in real time. We are only *perhaps* fortunate because, if it is correct, much of the world could experience the consequences of significantly colder conditions, with all that this entails. If incorrect, however, the warming continues. Thus, climate change can be a double-edged sword.

On the NOAA website, January 2011 was listed as the 17th warmest January on record, compared to some others that have been increasingly warm since the year 2000. However, it is far from the warmest of recent years. Therefore, only time will tell if this could actually represent a step *away* from the warming trend. Since 2010 showed quite an alarming spike in upward temperatures we cannot read too much into any short-term statistic, but must wait instead for the longer-term scenario to evolve.

Each side of the debate has much invested in proving that the Sun either does, or does not, have a direct bearing on the warming of the last 30–40 years. Many research papers and studies have been undertaken in search of an answer, *the* answer. Although much has been learned and determined, not enough has been convincingly shown to those taking the IPCC position that an alternative explanation might have been found. However, some researchers are increasingly confident that a larger solar-climate connection may be in the process of being uncovered. Regardless, judging from the breadth of recent papers, statements and articles on the topic, it is clear that a unanimous voice among researchers is still far from close to being a reality.

The Heart of the Controversy

The IPCC's present position is that all effective forcing factors prior to 1970 warming were natural, and the Sun was the primary driver. Critics of that position point to the large increases of anthropogenic carbon dioxide that occurred between 1940 and 1970, when, in fact the climate showed a decided cooling trend. The IPCC has attributed this to volcanic aerosols, however. Before, during, and since that time, CO_2 continued to increase almost in a uniform upward slope (EPA) [9]. However, although the IPCC has strenuously maintained its position, critics have challenged this assessment, maintaining that although the same average increase in anthropogenic CO_2 has been in play all along (at least, since 1900), only *now* has the IPCC found it necessary to implicate it! They claim this is simply because they have no other explanation for recent warming, despite what appears to be an inconsistent position.

The controversy regarding whether natural or manmade causes are responsible for recent climate change, then, lies at the heart of the entire debate. Meanwhile, although the Sun indeed might slowly have grown in irradiance between 1970 almost to the present (albeit with some leaner years), this still remains a subject of controversy. However, since 2001, most researchers seem to agree that overall solar activity has not increased at all, and the Sun actually might have entered a significant period of lesser activity, not only in sunspots (Fig. 4.6a) but also in the strength of its magnetic field (Fig. 4.6b).

Although readings of total solar irradiance and magnetic activity during the period do bear this out, the range of increases and decreases nevertheless is too small, say some, to have any direct warming effect [10]. This is, perhaps, the second key, however. If the Sun were to be, in fact, somehow responsible, then the warming has to be occurring from some indirect mechanism triggered by it. (Let us ignore cosmic ray scenarios for now.) Since the IPCC has only allowed for the direct warming effects of the Sun in their models, this gets to the core of the argument. Some researchers point to one possible explanation: extreme ultraviolet radiation (EUV) that might energize reactions in the high atmosphere, creating amplification of the warming. More about this later.

A curious position was taken by Alan. S. Brown in an otherwise extremely good article on the climate controversy [11]. In it he stated that IPCC models – which by and large he supported – did not *allow* for the direct effect on the climate by the variations of sunspot activity! It sounded a bit like Canute again, and certainly seems to be a case of putting the cart before the horse, since it is not the *models* that determine things.

The article is doubly curious, though, because Brown is partly critical of 'skeptics,' who he summed up as engineers and scientists, of whom only some have climate-related degrees. This might be true only in part, but maybe it is not true just of skeptics. Although he is apparently a well-educated individual in related fields, Brown, like many of the other skeptics he targets for criticism, also appears to have no credentials in climate science. Not that he should have to in order to have a thoroughly informed opinion, but he created a contradiction of sorts by his own validation of the IPCC. Regardless, he concluded with a very reasonable and perceptive remark that it is not only modelers but the "true skeptics" who will be those to advance our understanding of the climate.

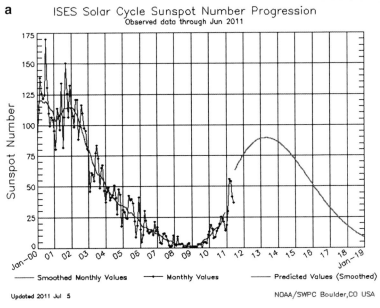

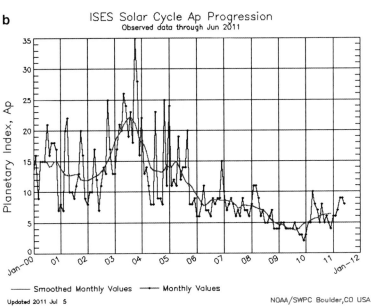

FIG. 4.6 (a) Sunspot activity, with projected cycle activity. (b) Solar Magnetic Planetary Index, Ap (solar activity at a planetary scale)

A Solar Driver?

Although some scientists today believe that the Sun may indeed be the primary driver of present-day climate change (see Gusev et al. and Hoyt and Schatten [7] this is in direct contradiction to today's predominant anthropogenic theories. Certainly, such alternate concepts are controversial within the mainstream, to say the least. However, those scientists engaged in researching these theories maintain there has been a virtual disregard of looking beyond the status quo. Regardless, there has been a wide variety of findings, ranging from confirmations of significant solar involvement to virtually none at all. Contradictions seem to be the norm in all areas of climate change studies.

The Sun's intense magnetic field (Fig. 4.7) is directly linked to many aspects of its activity. When the Sun is in an active phase we can expect consequences on Earth, especially with communications and transmissions of all kinds. This is because of the compression it causes to Earth's magnetosphere, as well as changes to the ionosphere – the tenuous atmospheric component termed 'near-Earth space.' With energized ions, increased radiation is able to penetrate normally protected regions and cause damage to satellite communication.

Fɪɢ. **4.7** The Sun's magnetic field and solar wind (Graphic image courtesy of NASA)

Solar flares are more likely to occur near irregular spots. Plasma loops are frequently precursors to flares; usually when sunspots exist in pairs or multiples, being magnetic opposites, plasma loops form bridges to their opposite counterpart. Figure 4.8a shows how this strong force is actually made up of multiple strands of plasma material bridging one sunspot to another; these strands together make up the more familiar loops. Figure 4.8b presents the more general appearance of loops at greater distances.

A Larger Solar Influence?

Recently, *National Geographic* made reference to research that had determined that the effects of solar variation were 'negligible.' However, it was added that other more complex mechanisms from increased solar irradiance might possibly play a role.

Such an acknowledgement of *any* other possibilities is rare to find in most similar mainstream sources.

Burroughs (see Chap. 2) pointed out that although sunspot cycles *also* hunt in pairs, they do so with alternate polarity from positive to negative, or vice versa. Although each spot will tend to have a companion of the opposite polarity, as a group, they alternate from leading to trailing each successive year. Sunspots also take opposite polarities on each side of the solar equator. Thus, with the annual alternations of polarity, we have a double cycle, which is better known as the 22-year 'Hale Cycle.' It is significant that this double cycle is usually more pronounced in climatic records than is the single 11-year cycle. The solar magnetic polarity controls the direction of the interplanetary magnetic field, which, in turn, reacts with the solar wind in Earth's magnetosphere. It has been argued that this could create the mechanism that amplifies the effect of any increases in solar irradiance, and therefore indirectly the climate.

Once it had been determined that sunspots appeared to follow an almost uniformly regular ~11-year cycle of growth and

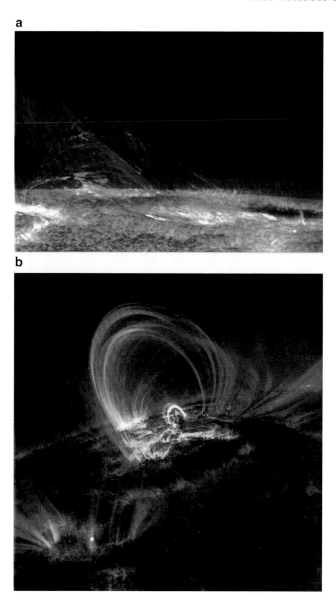

Fig. 4.8 Solar plasma loops ((a) Image courtesy of Hinode JAXA/NASA 2007. (b) Image courtesy of NASA)

decline (with the end of one cycle overlapping the beginning of the next), it was also noted that the number of sunspots within these sequences followed a pattern of especially fast growth and slower decline.

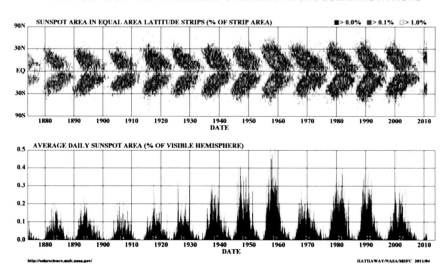

FIG. 4.9 Solar butterfly and relative sunspot surface area (Graphs courtesy of NASA)

Figure 4.9 consists of two paired graphs. The top graph is of a classic solar 'butterfly,' whereby the progressive growth of the sunspot appearances toward the solar equator can be seen in both hemispheres through each 11-year cycle; they echo each other almost perfectly. Although both graphs reference the same statistics, each one adds understanding to the other. On the lower graph, note the small amounts of sunspot cover relative to the total solar disc, despite the appearance of huge swings on the graph below. These phenomena cover no more than 0.5% of the entire solar surface at any time. However, far from decreasing solar output by that much, we can see that the potential net increase of 0.1% indicates that the cooler sunspots can account for as much as 0.6% of the activity.

In a 2003 NASA study it was estimated that solar irradiance had been increasing with each 11-year sunspot cycle at a rate of 0.5% since the nineteenth century [12]. Furthermore, one of its authors, Dr. Richard Willson (Principal Investigator of NASA's ACRIM solar satellite program), ventured the not insignificant opinion that if such a trend had indeed persisted throughout the twentieth century, it would have accounted for a "significant

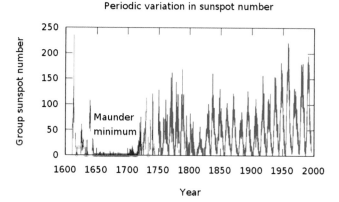

FIG. **4.10** Periodic variation in sunspot number (Graph based on data from Hoyt and Schatten [14])

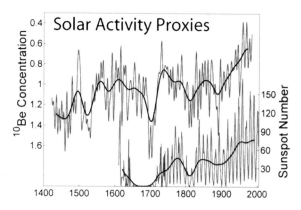

FIG. **4.11** Variable output of the Sun (Graphic by Robert A. Rohde, courtesy of Global Warming Art project)

component" of the warming of the past 100 years that had been featured in the IPCC reports.

Indeed, a gradually increasing trend of solar activity, as recorded through sunspots until the end of the twentieth century, can be noted on the 1997 graph in Fig. 4.10 (earlier, less reliable proxy readings are in red). The graph shows the sunspot minimum (Maunder Minimum) and seems to correspond clearly with the 'Little Ice Age' period. It should be noted that sunspot activity has had some mild downturns in the intervening years since that time (see paragraph following Fig. 4.11), but overall it climbed in intensity through the duration of the graph sampling.

Figure 4.11 illustrates some contrasting, but similar, evidence corresponding to such positions; regardless of the interpretations of these findings, it does indicate overall measurable increasing solar activity since the period of the Maunder Minimum and today. The blue portion of the graph references deposits of beryllium-10 in ice core samples. These deposits are directly related to solar magnetic activity (data from Beer et al. [13]). The red portion is based on historical observational records of sunspot activity (according to Hoyt and Schatten [14]).

The close correspondence of the upper points of activity in the ^{10}Be (beryllium isotope) proxy records and solar activity relative to sunspot numbers is striking. So, too, is the 11-year sunspot cycle, in shape, marching almost in lockstep from one set of data points to the other, but more specifically from century to century. Measurements since 1980 do appear to show some correlation to increased temperatures, in contrast to the fundamental position of IPCC positions that the warming of the last 30–40 years is almost entirely due to anthropogenic causes.

It is interesting to note that traces of other cycles may be glimpsed on these graphs. The so-called 'Gleissberg' 70- to 100-year cycle is one; however, more detailed graphs are needed to show that cycle definitively. It should be noted that the Gleissberg cycle encompasses more than just the number of sunspots, and includes cycle length and sunspot structure. In his 2000 paper, Shahinaz M. Yousef provided dates for the most recent activity of these cycles [15]. Minimums were indicated at 1797–1823, 1877–1913 and projected 1997–2032. Maximums were shown in 1778, 1860 and 1981. Although minimum periods of the cycle can be seen with a little effort in these graphs, not all aspects are completely clear, once again revealing how convoluted the statistics of climate change truly are. The last maximum in the series is also considerably longer than the norm, so it may be easy to miss on casual inspection of any graph.

The graph (Fig. 4.12) shows the degree that sunspot activity has varied over an 11,000-year time frame. The Sun shows itself as anything but monotonously constant, in spite of its stability compared to other stars. In an historical context of thousands of years, present day values do not appear remarkable.

In slight contrast, Solanki et al. had already come to the opinion that solar output in the last 70 years had been exceptional, the

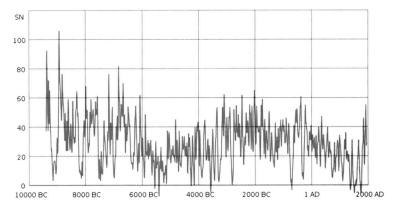

Fig. 4.12 Sunspot activity over the past 11,000 years (Graph by 'Conscious,' based on data from Solanki et al. [16], '11,000-year sunspot reconstruction' IGBP Pages/World Data Center for Paleoclimatology Data Contribution Series #2005-015. NOAA/NGDC)

present time corresponding only to the period 8,000 years ago [16]. They also found that solar activity of a similar level to that experienced since 1940 registered to the same degree for only 10% of the entire previous 11,000-year period. Proxy records, although imprecise, give us a better indication of what has occurred in the past, and what may be in store in the future. More accurate data of recent times will undoubtedly provide measurements by which theories of possible solar interactions with climate will be tested, especially with regard to the present rates of CO_2 increase. Should temperatures *fall*, in step with solar activity, or even soon thereafter, it would cause quite a stir in the climate research community.

Further reinforcement of the contention that there is an indirect link with solar variation and climate may be found in an article in *Science* [17]. It was remarked that a link of radiative forcing to the El Nino "ocean thermostat" could be demonstrated and confirmed by proxy records over the course of the Holocene. This does correspond with the findings of Kevin Trenberth, as discussed in Chap. 3, but the link to ENSO events is still not considered by all researchers to have been proven.

If the results of small changes to the Sun's radiant output are not immediately reflected on Earth with each cycle, a time delay between contributing solar activity and reaction of Earth's climate is a possible explanation. This has prompted several theories about such a possible phenomenon, termed Long Term Persistence, and its potential causes (see Chap. 6, [11, 17]). If the phenomenon exists, the present sunspot Cycle 24 might be slow to oblige in supplying evidence of a cooling trend, effectively allowing continued assertions on both sides about what is actually occurring in Earth's climate.

Opposing the notion of "majority opinion" of those favoring the IPCC position, some researchers believe that failures in existing climate models have resulted from not including all possible inputs, but actually from discounting some possible key factors. Only as recently as in the 2007 TAR would the IPCC make direct reference and acknowledgement of the 11-year solar cycle, as well as its effects on atmospheric ozone, the various hypotheses of the Sun's influence on cloud formation (including through direct interactions with cosmic rays), possible tropospheric changes associated with the cycle, and warmer, wetter periods at solar maximum, including winds in the upper atmosphere. It also referenced the difficulties of determining certainties in creating projections.

K. Georgieva, in his paper, 'Why the Sunspot Cycle is Double Peaked,' *Astronomy & Astrophysics*, Vol. 2011, Article ID 437838, remarked that many sunspot cycles show a double peak. As early as 1967, it had been thought that all cycles might have two peaks brought about by other processes, but the time between them often being too small to distinguish from measurements of the total sunspot activity.

More telling still was the remark commenting on "substantial uncertainty in the identification of climate response to solar cycle variations, because the satellite period is short relative to the solar cycle, and because the response is difficult to separate from internal climate variations and the response to volcanic eruptions." This certainly sounds like something of a partial concession to the hailstorms of critics who had maintained that some of these factors were being ignored. However, there does not appear to have been any mitigation to their overall position or models, in spite of the acknowledgment of so many uncertainties. But it does at least appear to be a beginning of a search for middle ground, uniformity and (hopefully) agreement.

Solar Radiation, Ultraviolet Light, Ozone and the Atmosphere

In 1980, new solar satellite surveillance provided consistent measurements of solar irradiance, including ultraviolet light transmissions at the high end of the spectrum. Because the solar flares (related to sunspot activity) were thought likely to be implicated in increased ultraviolet emissions, it was important to determine what effect they had on Earth's environment. In 1990, a carefully crafted model by noted researchers Judith Lean and Peter Foukal detailed how ultraviolet radiation measurements now could be calculated back to the beginning of the twentieth century, and even earlier [18].

The Effect of Solar Variation on Climate

In the Stanford Solar Center (online), it was stated that a "recent review paper" by both solar and climate scientists had concluded that although the Sun might be implicated in climate change to a small degree, its effect was much smaller than estimates of forcing from anthropogenic contributions. In other words human activities were held to blame as the primary factor in global climate change.

Here we have another example of a blanket statement made in the absence of citing the actual study referenced, and from a major institution, no less. Whenever we become aware of such positions stated in a vacuum, we should always suspect that there might be other arguments from the opposite side of the coin.

Ultraviolet radiation's relationship with Earth's atmosphere is complex and also varies disproportionately with solar activity, with small increases in activity producing large increases in high-frequency UV. Cornelius de Jager ventured the opinion that interactions between solar plasma ejections with cosmic rays, together with UV irradiance due to variations during the 11-year cycle, are among the 'prime suspects' in climate change. An exhaustive analysis of the research into his position may be found in his 2008 paper [8]. In another paper dating from 2006, he and Ilya Usoskin concluded that variations in solar UV flux are likely to affect temperatures more than variations in cloud cover from cosmic rays (for more on cosmic rays and Earth's atmosphere, see also Chap. 12) [19]. Burroughs also dealt comprehensively with the topic of UV, cosmic rays and climate in his book.

Photochemical processes also affect the composition of the atmosphere. Oxygen (O_2) absorbs solar radiation under 240 nm, but UV produces more energetic auroral and proton particles that break apart O_2 into atomic oxygen (O), a free radical. Other free radicals are created, too, including nitric oxide (NO), hydroxyl (OH). Ozone (O_3) is produced as a consequence of ultraviolet radiation photochemical/catalytic reactions with other free radicals, while this double ionization of oxygen (O_2) and atomic oxygen (O) also warms the surrounding air. Ozone also absorbs solar radiation. UV is readily absorbed by ozone from 240 nm up to 310 nm in the upper stratosphere, providing protection below on Earth against electromagnetic radiation.

Chlorofluorocarbons (CFC's) are organic compounds that contain carbon, chlorine, and fluorine, volatile derivatives of methane and ethane. Commonly related subclasses are hydrochlorofluorocarbons (HCFC's) with hydrogen partly comprising them. The refrigerant dichlorodifluoromethane (R-12 or Freon-12) is one of the most common examples. Others may be found as aerosol propellants and solvents.

However, like atomic oxygen (O), ozone (O_3) is a very unstable gas, soon to be stripped of an atom by the absorption of solar radiation (!) to become once again oxygen and atomic oxygen, repeating the cycle in a seemingly endless loop. Additionally, man-made chlorofluorocarbons (CFC's) have been implicated in playing a role. Because there is a natural balance of ozone depletion and production high in the stratosphere, the addition of chlorine (Cl) from the photochemical reactions with CFC's, along with atomic oxygen (O) can also act as a catalyst in destroying ozone.

Some researchers seized upon this concept as possibly the Holy Grail. Krikova et al. concluded that although solar irradiance is the main external driver of climate, it drives it through the effect of the UV component on the chemical and physical processes of the upper atmosphere [20]. They argued that this factor was especially significant, since solar irradiance had increased overall during the past four centuries.

A characteristic of increased ozone is incrementally lower transmission to Earth's surface of the solar spectrum at increasing latitudes, because of the greater diffraction relative to lower concentrations. The effect of this on climate becomes most significant during wintertime during cycles of increased solar activity. It has been theorized that increased relative warming of layers of the middle atmospheric, in turn, creates significantly wider temperature differentials between the poles and equator, with resulting greater Hadley cell convection processes. Thus, such altered air circulation currents may be responsible for more unstable conditions incrementally towards the poles, and with that, the greater the likelihood of colder, more inclement winters in the higher latitudes during these times.

It should not be surprising, however, that we find reduced ozone towards the poles (the so-called 'ozone holes'), because it is here that UV is less prevalent due the increasing angles of the incidence of solar radiation, and CFC's destructive role is undiminished. We can also see how Earth's seasons may play a role in ozone concentrations through changes in the amount of UV reaching the upper atmosphere. It is not clear, however, that anyone at this time is able to determine clear specifics of such complex processes, let alone predict their effect adequately for any given climate model or scenario to be stated with precision.

U. Langematz et al. presented an analysis of these processes in detail, as well as the complex resulting effects of atmospheric interactions [21]. Significantly, her group's concluding remarks stressed that climate models needed to reflect the input of solar variations, together with better observational data, in order to attain accurate assessments of the Sun's influence. (The extensive volume of studies featuring her co-author, K. Matthes, and those related to the 11-year cycle, atmosphere, ocean currents, ozone, geomagnetic forcing, etc., is also a highly interesting resource [22]).

Lean's and Foukal's names remain permanently associated with solar research and its relationship with Earth's climate. Lean's follow-up study in 1997 to her 1990 landmark paper with Foukal continued to stress the importance of solar activity on Earth's climate, although she was not prepared to state that definite links had been established [23]. She found the link to depletion and contributions of anthropogenic CFC gases within the 11-year sunspot cycle to be striking, and determined that the effects of EUV (extreme ultraviolet) radiation on the upper atmosphere were "undisputed."

Extreme ultraviolet radiation (EUV) is a high-energy ultraviolet radiation, spanning wavelengths from 120 nm down to 10 nm. EUV is naturally generated by the solar corona and artificially by plasma and synchrotron light sources (radially accelerated to higher frequencies, such as magnetically).

We should also bear in mind that since the Sun's energy spectrum is not evenly distributed, and increased radiation is disproportionately skewed towards the higher frequencies (i.e., UV and EUV), this may be another key. Lean also confirmed that solar activity was at an historically elevated level. Stressing the uncertainty of determining the degree that the Sun's influence is affected by fractions of a percent of forcing changes, Lean further remarked that the many unknown and likely factors in the twenty-first century would complicate people's abilities to determine the total natural and anthropogenic forcings and feedbacks (positive or negative) of future climate projections.

Interestingly, Lean acknowledged that there was "cautious confidence" that the 11-year solar cycles do indeed affect variations in temperature, ozone, and winds. This was a striking position compared to some of the heated debate that has occurred in the interim. There were also interesting comments about the problems facing climate modelers, instead of just insisting that they had included all presently known factors.

All in all, this report makes an impressive and balanced assessment of as many factors as could be accounted for at the time that could be realistically theorized or demonstrated. However, in a 1996 article ('The Sun and Climate') located on the website of the U.S. Global Change Research Information Office, Lean's curious and repeated use of the terminology 'burning' in describing the workings of the Sun is hard to reconcile with a scientist of her reputation, or indeed, the U.S. Global Change Research Information Office where the article may be found! [24].

Looking for Common Ground

If looking for common ground sounds political and not what we should be contemplating, at least it speaks to reason and calm debate. Noted Israeli researcher Nir J. Shaviv provides just such a demonstration of reason in his discussion on his website Science-Bits, where he contemplates that the truth may lie somewhere in the middle – natural forcings being the dominant driver in the twentieth century, and anthropogenic forcing assuming increasing significance in the twenty-first [25]. However, he does state that there is no evidence for the link of greenhouse gases to the observed increasing temperatures, only that they are a possible factor and a theoretical forcing agent. Furthermore, he states that carbon dioxide has only been implicated because of the lack of another explanation, something we have heard before. It is also interesting that he blames the media for the simplistic picture painted of the mechanisms involved in climate change, as well as many climate scientists. Strong words indeed.

Ultimately, Shaviv looks to indirect solar forcing, whereby climate warming results from secondary responses to the initial solar forcing, while he maintains that we still have no way to measure the extent of anthropogenic forcing. We will return later in this book to some of Shaviv's own hypotheses, which look to astronomical causes for some of the explanations. A 2001 paper by Pål Brekke, effectively drew a similar conclusion, that both anthropogenic and natural causes could play equal or dominant roles in climate change [26]. Significantly, he pointed to the weakness of existing climate models, because of the input of direct forcing effects of solar irradiance, a sentiment that also has been expressed frequently and repeatedly by others.

Other research also seems to have its feet firmly planted in both camps. The authors Natalia G. Andronova and Michael E. Schlesinger saw truth in all viewpoints, although they concluded anthropogenic forcings were the present primary drivers [27]. However, they entertained the possibility, presented in 1997 by no less a figure than James Hansen himself (director of the Goddard Institute for Space Studies, and perhaps the most renowned of all AGW proponents), that unknown forcings might also be responsible. Apparently Hansen no longer believes this, arriving at a place where he and others have parted ways.

References

1. Svensmark H (2007) Cosmoclimatology: a new theory emerges. Astron Geophys 48(1):1.18–1.24
2. Svensmark H, Christensen EF (1997) Variation of cosmic ray flux and global cloud coverage – a missing link in solar-climate relationships. J Atmos Sol Terres Phys 59(11):1225–1232
3. Clark S (2010) What's wrong with the sun? New Sci (2764)
4. Archibald DC (2006) Solar cycles 24 and 25 and predicted climate response. Energy Environ 17(1):29–36
5. Niroma T (2009) Sunspots: from basic to supercycles. Website page: http://personal.inet.fi/tiede/tilmari/sunspot4.html
6. Abdussamatov HI (2004) About the long-term coordinated variations of the activity, radius, total irradiance of the Sun and the earth's climate. Inter Astro Union 541–542
7. Gusev A, Pugacheva G, Inácio MM, Spjeldvik W (2010) Climatic variability as natural oscillation driven by solar activity variations. In: 38th COSPAR Scientific Assembly, Bremen, Germany; Hoyt DV, Schatten KH (1997) The role of the Sun in climate change. Oxford University Press, New York
8. de Jager C (2005) Solar forcing of climate. 1: Solar variability. Space Sci Rev 120:197–241
9. EPA. Climate change science: atmosphere changes. Website page online: http://www.epa.gov/climatechange/science/recentac.html

10. Conn TA, Lins F (2005) Nature's style: naturally trendy. Geophys Res Lett 32:L23402
11. Brown AS (2010) Climate models and their critics. The Bent of Tau Beta Pi, www.tbp.org/pages/publications/Bent/Features/W10Brown.pdf
12. Willson RC, Mordinonov AV (2003) Secular total solar irradiance trend during solar cycles 21–23. Geophys Res Lett 30(5)
13. Beer J, Foos F, Lukasczyk C, Mende W, Rodriguez J, Siegenthaler U, Stellmacher R (1994) [10]Be as an indicator of solar variability and climate. In: The solar engine and its influence on terrestrial atmosphere and climate. Springer, New York
14. Hoyt DV, Schatten HK (1997) Group sunspot numbers: a new solar activity reconstruction, parts 1 and 2. Sol Phys 179:189
15. Yousef SM (2000) The solar Wolf-Gleissberg cycle and its influence on the Earth. In: Proceedings of the international conference on the environmental hazards mitigation, Cairo University, Cairo
16. Solanki SK, Usoskin IG, Kromer B, Schüssler M, Beer J (2004) Unusual activity of the sun during the recent decades compared to the previous 11,000 years. Nature 431(7012):1084–1087
17. Marchitto TM, Muscheler R, Ortiz JD, Carriquiry JD, van Geen A (2010) Dynamical response of the tropical Pacific Ocean to solar forcing during the early holocene. Science 330(6009):1378–1381
18. Foukal P, Lean J (1990) An empirical model of total solar irradiance variations between 1874 and 1988. Science 274(4942):556–558
19. de Jager C, Usoskin IG (2006) On possible drivers of Sun-induced climate changes. J Atmos Terres Phys 68:2053
20. Krikova NA, Vieira LEA, Solanki SK (2010) Reconstruction of solar spectral irradiance since the Maunder minimum. J Geophys Res 115(12)
21. Matthes LK, Grenfell JL (2005) Solar impact climate: modeling the coupling between the middle and the lower atmosphere. Soc Astron Ital
22. List of publications for K. Matthes: Nathan.gfz-potsdam.de/images/Publications_KatjaMatthes_Nov2010.pdf
23. Lean J (1997) The Sun's variable radiation and its relevance for Earth. Annu Rev Astron Astrophys 35:33–67
24. Lean J (1996) The Sun and climate. Article found on the website of the U.S. Global Change Research Information Office. http://www.gcrio.org/CONSEQUENCES/winter96/sunclimate.html
25. Shaviv NJ (2010) Carbon dioxide or solar forcing? ScienceBits. www.sciencebits.com/CO2orSolar

26. Brekke P (2001) The Sun's role in climate change. Presentation at the 1st international conference on global warming and the next ice age TH 1.2 p. 245
27. Andronova NG, Schlesinger ME (2000) Causes of global temperature changes during the 19th and 20th centuries. Geophys Res Lett 27(14):2137–2140

5. Short-Term Climate Variation

Climate Change in the Past Thousand Years

Much has been made of possible variations in the climate over the past 1,000 years, with strong arguments presented on both sides. Such variations in climate have been indicated in many types of historical (proxy) records, and to some degree reported in literature, with extended periods theorized as substantially warmer and colder than those of today. However, some researchers continue to question exactly what took place, if indeed, anything did at all. This is simply because of the lack of consistent and reliable records; proxy data is all we have.

Proxy Evidence for the Warm and Cool Periods

Proxy evidence – indirect indicators of the historic record – is deduced from many sources. These records are wide in scope, ranging from extrapolating power spectra relative to solar activity, or the state of human development from known factors including carbon-14 and oxygen-16/18 dating of ice cores, geologic evidence of climate variations, sea level records, sea ice variations over the past 9,000 years with new sediment analyses of its effects on deep ocean currents, utilizing argon and nitrogen isotope records, even research in the Sargasso Sea. By such means many researchers believe they have confirmed climate variations of at least 1°C in both directions [1].

Proxy evidence from the past 1,000 years points to a warm period in northern Europe, based on indirect dating techniques. Although many supporters of the IPCC position have sought to question the reality of the warm period (because it lacks the drivers of warming utilized in present-day climate models), proxy evidence includes glaciers that show traces of vegetation at their

A. Cooke, *Astronomy and the Climate Crisis*,
Astronomers' Universe, DOI 10.1007/978-1-4614-4608-8_5,
© Springer Science+Business Media New York 2012

extremities dating from *after* that time – a sure sign of recent renewed glacial expansion. There are numerous proxy indicators of the cold period at many locations worldwide, too, although they do not seem to have occurred necessarily at precisely parallel times.

The evidentiary trail has been sufficient for some researchers to believe there is nothing out of the ordinary in conditions at the present time of writing. However, this trail, though not conclusive, points to significant periods within the past 1,000 years that do not equate with what we would consider the predominant climate of the Holocene, especially to the most notoriously controversial supposedly "warm" and "cold" periods (the Medieval Climate Optimum and the Little Ice Age). It is hard to determine if the Medieval Climate Optimum might have been comparable to the warming experienced during the course of the twentieth century. If, as some have proposed, the Sun's small variations are sufficient to affect climate noticeably, such a finding would be significant indeed.

What is clearly significant now, however, is that some of the more recent proxy research stands in direct contradiction to some of the older studies, which in the meantime have become staples of IPCC protocol [2]. Following measured acceptance of both phenomena by the IPCC in its first assessment in 1990, the panel gradually moved towards rejection and discredit of the Medieval Climate Optimum and the Little Ice Age due to much of the older research. Thus, in just over 10 years, the IPCC moved from its original position to the opposite (1990 FAR to 2001 TAR). In doing so, the phenomena, clearly present in 1990, somehow had vanished without a trace by 2001. These changes of direction did not go unnoticed, as critics of the IPCC were quick to pounce and accuse the organization and those who supported it of changing their views to suit the official stance.

The Medieval Climate Optimum

The concept of warm and cold periods is hardly new. Most notably, it had been raised by other earlier twentieth century researchers in 1965 by H. H. Lamb, the iconic climatologist. His writing, "The early medieval warm epoch and its sequel," remains monumental to this day, although not without controversy [3]. Researching proxy evidence from the Southern Hemisphere to the arctic north,

Lamb believed that approximately 1,000 years ago temperatures were as much as 2°C higher than those of the cooler mid twentieth century decades prior to the 1960s, even up to twice that in higher latitudes. Interestingly, phases of human development seem to have coincided with such proposed periods, in that warmer periods brought about prosperity and advancement of civilizations, whereas cold times effectively stalled it [4].

During the other period in which it has been theorized the climate was warmer than any time up to the present – the Medieval Climate Optimum – some have claimed temperatures rose well in excess of anything we have experienced in recent times. Certainly there are no eyewitness records of sunspot activity during medieval times, so it is not possible to show increased solar output by direct records. Many researchers have questioned the reliability of the only indicators we have – proxy records.

The Medieval Climate Optimum

Far from accepting the premise of the Medieval Climate Optimum, many commentators claim that other regions of the world became cooler when Greenland warmed in medieval times, and also that the average global temperature today is higher than it was during that period.

Regardless, any commentary that states as a matter of fact things that remain controversial, or unclear at best, are not helpful. In truth, we really do not know exactly what occurred during medieval times and can only hope that one day it will be possible to end the debate; meanwhile, a lot depends on whom one's faith rests.

Some scientists have proposed that the mechanisms responsible for present-day warming are no different from those that occurred during medieval times, and that such climate variations are natural and recurring. Those who have claimed that the occurrence of these warm and cold periods is questionable have found it more difficult to justify how some of the artifacts from medieval times came into being. Without a warmer climate playing a role, what could have caused the conditions necessary to enable their existence is still undetermined.

The Little Ice Age

Historically, if we accept the premise of proxy records, in the past 1,000 years we can see that a period of minimal sunspot activity (the Maunder Minimum) appears to correspond with the reported coldest portion in the middle of an apparently larger cold period in the climate (1550–1850), a time better known as the Little Ice Age; the later Dalton Minimum also seems to fit the latter part of that overall period. (The Maunder and Dalton Minimum may be seen on the graph in Chap. 4, Fig. 4.11; on the graph, the unnamed Dalton Minimum is the next significant dip in solar activity that occurred after the named Maunder Minimum.)

The Maunder Minimum and the Little Ice Age

The period of solar inactivity (the Maunder Minimum) also corresponds to a climatic period called the Little Ice Age when rivers that are normally ice-free froze and snow fields remained year-round at lower altitudes. There is evidence that the Sun has had similar periods of inactivity in the more distant past. The connection between solar activity and terrestrial climate is an area of ongoing research.

–NASA

NASA does seem to have taken the position that the Little Ice Age did, in fact, exist, remarking on its website, "During the Little Ice Age from the early 1400s to late 1800s, the vegetation changed again to plants that favored cooler and wetter climates. The core records revealed increases in spruce and hemlock that prefer cooler and wetter climates."

Although not a true ice age according to precise definitions, the Little Ice Age nevertheless appears to have been an extended period of decidedly cold conditions, especially significant to the advancing civilizations of northern Europe. A chilly impression of these times seems to be maintained throughout literature (i.e., in Charles Dickens' England) as well as in art (see Fig. 5.1), right

Fig. 5.1 *Winter landscape with iceskaters,* c. 1608; Skating on a frozen river during the little ice age (Oil painting by Hendrick Avercamp)

up to the end of the first half of the nineteenth century. Life would have been very challenging – even more so than usual – if a true cold period was something the populace of the time had to deal with.

However, changing patterns in the way of life of the time, the varying dates of the apparent occurrence of the cold period around the world, as well as the state of early technology, are among the difficulties in determining exactly what took place. Because disputes still rage about solar activity, or more precisely, the lack of it, knowing exactly *how much* colder (if at all) the Little Ice Age actually might have been could be pivotal. Research has included even the study of small changes of Earth's atmospheric transparency to solar radiation that might have proved sufficient to trigger cold periods or ice ages [5]. Certainly the indications are there: contemporary depictions of skating on frozen rivers, "winter ice fairs" and "ice carnivals," etc. However, simpler explanations might provide at least some of the answers. For example, different uses of waterways may have contributed sufficiently to water stagnation to allow it to freeze; it has been argued quite reasonably that later dredging of riverbeds could have greatly retarded ice formation.

Trying to Explain Both Periods

As recently as 2009, while agreeing that parts of the world experienced warmer times that corresponded to the Medieval Climate Optimum, the prominent climatologist Michael Mann was still clinging to his conviction that the warm period was not a global phenomenon at all, but strictly localized in relation to attributes other than true drivers of climate. In two 2002 encyclopedia articles, although conceding that the cold period had existed in northern Europe, Mann raised the possibility that volcanic activity or the North Atlantic Oscillation could have resulted in cooling taking place in *that* location, but not globally [6]. In his parallel entry, Mann related similar views, while acknowledging that indeed the climate in high latitudes of Europe appeared to have been warmer during earlier times than several centuries later. He indicated that in some regions of the Northern Hemisphere the range of temperatures might have been more pronounced than in others, although in general they remained largely constant.

Overall, Mann remained guarded about pronouncing anything out of the ordinary for either period, citing lack of supporting evidence, viewing, for instance, the demise of the oft-quoted settlements in Greenland as being due more to declining socioeconomic factors than a cooling climate. Proxy evidence in Greenland (even more revealed by its name) shows Viking occupation and agriculture, and much of the landmass they occupied (including gravesites) apparently remained deeply frozen for centuries since. Although Mann agreed with the view that Norse setters had inhabited the southernmost coastal portion of the country in medieval times, this was as far as he was prepared to go. Interestingly, also in the second of the two articles, Mann raised the possibility of astronomical factors coming into play and contributing to the Little Ice Age.

General estimates for potential warming since the Maunder Minimum has been widely studied; two examples out of many might provide some insights about some of the conclusions and methodologies used in their determinations [7]. Regardless of the cause, the warming during the past 100 years seems not to have followed the same pattern over the 1,000 years relative to historically projected solar variation.

Mann's "Hockey Stick"

In 1998/1999 Michael Mann headed the group that famously produced the original climate model for the past 1,000 years, better known today as the iconic Mann's Hockey Stick (Fig. 5.2) [8]. This showed no significant periods of early warming at all but did lead to a radical spike upwards into the new century beyond anything experienced during the entire sample period. Certainly 1,000 years ago, there was no evidence for elevated levels of CO_2 in the atmosphere.

Mann's graph caused an immediate outcry, because to many other researchers at the time, the numerous types of proxy records seemed to show things differently. However, the graph was a "star" feature in the 2001 IPCC Third Assessment Report (TAR), and thus became unwittingly perhaps the best-known climate change symbol for those who challenged anthropogenic global warming (AGW) theories.

Mann would concede that proxy records do contain "error bars," especially the further back in time from which they are utilized. Overall, however, most other independent research papers essentially came out in agreement. Perhaps Mann "goosed" the

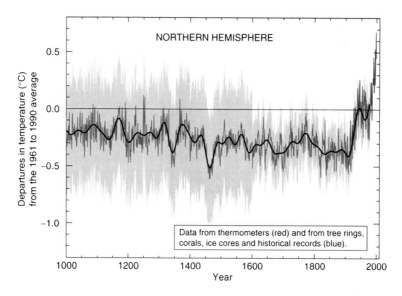

FIG. 5.2 Mann's "Hockey Stick" (Graph courtesy of IPCC TAR 2001)

argument by declaring that anthropogenic greenhouse gases were entirely responsible for the warming of recent past decades. Though clearly something he believed, Mann may not have realized it would set off a firestorm. He later admitted to regretting the way the IPCC had used his graph at the outset of their assessment to create the most sensational impact possible, because he considered it to be a work in progress.

Perhaps the most obvious shortcoming of the graph is its use by other researchers to "prove" the "Hockey Stick's" relationship of CO_2 to temperature. Some argued that if the situation were as dire as some made it appear, one might have expected global temperatures to already be far greater than at any time since the end of the last major ice age. Regardless, in 2006, the National Academy of Sciences concurred with the findings of Mann, which would be sure to keep the "Hockey Stick" firmly in view. Interestingly, unexplained discrepancies have been found to exist between the lag of observed temperatures in recent decades and changes in solar irradiance [9].

An Infamous Challenge to the "Hockey Stick"

One of the best-known and similarly hot-button studies of climate over the past millennia was that headed by Willie Soon, which examined the entire period from medieval times to the present [10]. It also has the dubious distinction for having stirred heated accusations of questionable peer-reviewing practices in the journal that published its first version *(Climate Change)*. In the paper, which was expanded, rewritten and later published in *Energy and Environment*, the researchers maintained that present-day warming was not exceptional. The authors examined a wide field of evidence, which was, by default, all proxy. A degree of confidence in each type of evidence was accorded, depending on the climate measurements deduced.

Soon believed the methods used by Mann were out of step with proxy climate evaluations, and that the use of empirical orthogonal functions (EOF's) to evaluate proxy readings had introduced a weighting towards the incorrect conclusions. Thus it was stated that Mann's study was biased too greatly on a worldwide model instead of factors that would apply to individual local situations. One does not leave the findings without the conviction that they were aimed primarily at trying to refute the "Hockey Stick" – an effort to "head it off at the pass." Mann rejected the study outright, saying its authors "got just about everything wrong."

But did Soon indeed get everything wrong? Certainly there is sufficient tree-ring evidence, for example, to justify claims of the medieval warm period from as far away even as New Zealand, certainly a great distance from northern European regions, and in a different hemisphere at that [11]. Soon had already proposed that such conditions were not limited to northern regions, but were in fact a far wider-reaching phenomenon. Interestingly, Soon also tied recent climate warming to ENSO events; with extensive worldwide evidence of the Little Ice Age, the hypothesis was that twentieth-century warming might be the late and final resolution of it. In essence he claimed to have produced an empirically based study versus a mathematical one. Significantly, he did not deny the possible influence of humankind on the climate historically, either, resting his case to a large degree on the recent strong warming trends not in any way synchronized to the near monotonic increases in CO_2 concentrations during the twentieth century. This is an observation we have already seen made by others, of course.

However, for those who claim that Soon was biased against theories of anthropogenic causes of warming, this is indeed mostly true, *mostly* because their position has been that the possible warming effects of carbon dioxide are possible but not proven. Regrettably, though, Soon's study has remained so controversial that it is hard to accept it without some recognition of its somewhat shaky position within the larger scientific community.

Compromise?

In 2005, a new ray of hope appeared, bridging the gap between the extremes of Mann and Soon. Better known as "Moberg et al.," it seemed to blend possibilities from both extremes into something likely to be a more accurate picture of the past 1,000 years [12].

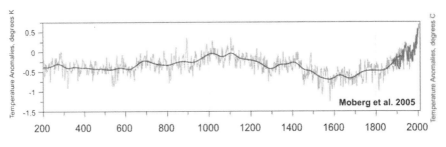

FIG. 5.3 Moberg et al. reconstruction (Courtesy of NOAA)

Still showing something of the "hockey stick" upward ramp at the end, it appeared less extreme, while the Medieval Warm Period and Little Ice Age became clearly discernable (Fig. 5.3). The authors, significantly, allowed for natural causes as significant contributors to the changing climate, including solar variation and aerosols. They did not consider present warming was due exclusively to natural or manmade causes, a significant departure from Mann's position.

The Question of the Sun's Involvement

The Sun has featured large in much of this recent research, often with conclusions that possible periods of climate change were most likely closely tied to variations of solar irradiance. Some studies have tied as much as two thirds of climate temperature fluctuations possibly to solar forcing and volcanic aerosols. This would seem to be in partial agreement both with Soon *and* Mann. Others have concluded, too, that these anomalies have caused too many problems to justify current climate models.

Studies of the Sun's possible role in climate are far from new. Although examples exist from the mid-twentieth century, there had always remained an inability to measure accurately the Sun's output, simply because of the limited history of solar observations and the technical means to measure it. Thus it has been difficult to compare the "active" and "quiet" Sun over the greater historical record.

It is very unlikely that the twentieth-century warming can be explained by natural causes. The late twentieth century has been unusually warm. Palaeoclimatic reconstructions show that the second half of the twentieth century was likely the warmest 50-year period in the Northern Hemisphere in the last 1,300 years.

–IPCC, TAR 2007

Some dramatic theories have emerged recently, in which the "quietest area" on the present Sun was compared against the minimum state of the historic Sun (according to isotope proxy data) [13]. One claimed that on time scales of 50–1,000 years the Sun had been the principal driver of climate throughout the past 7,000 years [14]. Unfortunately, any conclusions based on the recent past (from 1980) could not be applied, because newly available satellite data confirmed that the Sun was already in a maximum phase of output during this time. Even more regrettably, the focus of the climate debate today concerns the very years since then!

That the research providing this conclusion was not accepted for publication immediately upon completion made it the target of claims of irrelevance by many on the opposite side of the argument, even though it would appear soon after in *Astronomy and Astrophysics*. Some other well-written and well-researched papers similarly have had to wait many months for acceptance and publication. Again, this illustrates that not all members of the larger scientific community necessarily are benevolent to each other, especially on this hot-button issue.

Other solar research does indicate that we *might* be at the early stages of the downside of a recurring climate cycle, and that we can expect overall cooling until later this century [15]. A cooling of as much as 1.5°C by 2020 has been theorized, based on weak upcoming Solar Cycles 24 and 25. These cycles are regarded by some as similar to the Dalton Minimum, a period of weak activity that occurred between 1790 and 1830. However, it is too soon to know what is actually going to happen; hopefully we will have greater certainty over the next 10 or 20 years.

On the opposite side of the coin, some also have claimed the Sun is "not guilty" of involvement in recent changes to the climate, but for different reasons. In 2009, *Science Daily* led with a blanket headline about new findings that had concluded that solar changes were *not* responsible for global warming [16]. However, it turned out to be mostly centered on the potential of cosmic rays to trigger cloud formation via solar modifications to the rays (see Chap. 12) rather than the more immediate links we have been discussing. It seems that not everyone is talking about the same things! Thus it is possible to see that the range of concepts, from the hypothetical to the less esoteric, is wide indeed, though the Sun continues to be a prime focus of investigation.

The ACRIM Controversy

This infamous controversy best sums up all that has been at the heart of many of the arguments aired in this chapter – that is, the variability of the Sun's output and the potential effects of it. To get to the crux of it, we need to look at the satellite record from 1980, because it is only since then that we have had the advantage of accurate solar irradiance measurements beyond the atmosphere. Not only has it provided the specific measurements lacking before, but it is highly illustrative to realize that even with such tools, many of the more significant disputes from within the scientific community have not been quelled at all.

A NASA satellite program known as ACRIM is an ongoing space-based study to measure solar irradiance and its variations. Because earlier measurements (from balloons) had never been of sufficient reliability to make definitive judgments in the past, here at last was an opportunity to have the answers all climatologists had long been looking for. There have been five satellites to date involved in this program, the intent being to provide a consistent and continually monitored source of total solar irradiance (TSI), and thus finally to remove all guesswork. In essence, ACRIM promised to provide potentially a large part of – if not the entire – "smoking gun" that some climatologists had long been looking for. Perhaps unsurprisingly, it was only a matter of time until a major rift emerged over the findings.

First we should know something of the program. ACRIM-1, ACRIM-2, Nimbus-7/ERB and ERBS, and now ACRIM-3 were names given to Earth orbit solar satellite programs. It was considered that, out of them, the ACRIM satellites were the most sophisticated. However, a most unfortunate incident resulted in a period of 2 years elapsing between the operational status of two of these key satellites – ACRIM-2 having been delayed due to the failure of the space shuttle *Challenger.*

Critically, thus came about the now infamous ACRIM Gap, the 2-year period coinciding with a time when the NIMBUS-7/ ERB/ERBS programs were expected to become less reliable as they were reaching the ends of their operational life expectancy. Specifically, the better of the two earlier satellites (ACRIM-1) was out of action by the late 1980s, leaving the aging and slowly failing NIMBUS-7 to perform sole monitoring for a time. The ACRIM team therefore had to take into consideration a less than ideal scenario in arriving at its findings, an approach excerpted from an article on NASA's website, dated March 20, 2003:

> Richard Willson, Principal Investigator of NASA's ACRIM experiments, compiled a TSI record of over 24 years by carefully piecing together the overlapping records. In order to construct a long-term dataset, he needed to bridge a two-year gap (1989 to 1991) between ACRIM-1 and ACRIM-2.
>
> Both the Nimbus-7/ERB and ERBS measurements overlapped the ACRIM "gap." Using Nimbus-7/ERB results produced a 0.05 percent per decade upward trend between solar minima, while ERBS results produced no trend. Willson has identified specific errors in the ERBS data responsible for the difference.
>
> The accurate long-term dataset, therefore, shows a significant positive trend (.05 percent per decade) in TSI between the solar minima of solar cycles 21–23 (1978 to present). This major finding may help climatologists to distinguish between solar and man-made influences on climate.

However, this did not satisfy everyone. The ACRIM Gap would thus become the subject of such a hugely disputed component of the program that it almost derailed it. Another climate team (PMOD in Switzerland), led by researchers Claus Fröhlich and Judith Lean, questioned these findings and claimed that the TSI (Total Solar Irradiance) needed instead correction *downwards* because of the ACRIM Gap.

It is significant that the small adjustment by the PMOD researchers of the NIMBUS-7/ERB/ERBS data in question created an entirely different outcome from the conclusions of the ACRIM team. In two separate assessments (1998 and updated in 2006) Fröhlich demonstrated that solar irradiance had, in fact, decreased rather than increased. This resulted in a scientific feud that has become notorious, with Fröhlich's most recent response coming as late as 2008 [17]. Thus we have a relative minutia of the total data that resulted in two entirely opposite conclusions. The most sophisticated measuring system of all was unable to settle the argument, and thus, the potential silver bullet of the climate change debate had become the subject of the debate itself!

In response, Willson protested [18]:

- "The unauthorized and incorrect" adjustments made by PMOD researchers, "without any detailed knowledge of ACRIM-1 instrument or on-orbit performance, original analysis or any consultation with the ACRIM team."
- PMOD's "inferior and less accurate measuring systems."
- That PMOD had "pandered" to those who support anthropogenic causes of warming, a clear shot across the bow of the IPCC.

Meanwhile, an independent assessment of the controversial satellite readings was carried out at the National Solar Observatory in Kitt Peak, Arizona; records of the full solar disk taken at the observatory did, in fact, precisely confirm the ACRIM records. Since then, other more sophisticated satellites have been performing the monitoring, typically with considerable overlap, so their results are not in question. In fact they are quite compatible, and not the subject of any dispute. Thus it comes down to two critical years!

Solar Activity

In efforts to support the AGW position, we frequently hear claims that recent global warming and solar activity have been moving in opposite directions.

Although this position is indeed supported by the PMOD position, attempts to take it preemptively, in lieu of any acknowledgement of the alternate conclusions of the ACRIM team (responsible for the satellite program itself that has measured solar activity since 1980) may not be noticed by most people. The ACRIM conclusions, polar-opposite in virtually every way, indicated an *increased* level of solar activity until the beginning of the twenty-first century. In any fair and unbiased presentation of information, readers are entitled to full disclosure. They will not necessarily receive it from all sources, and ought thus to remain wary.

We can examine the ACRIM researchers' findings from the graphs themselves, having become some of the most widely known of all. To make sense of what may appear to be complex, first we must recognize, overall, the 11-year solar cycles, shown here in various segments (according to the satellite that had provided the data) as waveforms representing peak and minimum output, up and down, like hills and dales. The five satellites in use for the program had different and varying life spans, none of them covering the entire period represented over the entire graph.

The period of each solar cycle can be referenced along the top of the graph, showing Cycles 21–24. These correspond to the dates along the bottom of the graph. If we attempt in our minds to merge all the satellite solar output waveforms (each one named) into one horizontal level (i.e., raising the red and black waves all to that of the blue level), it will result in a single up and down curved horizontal waveform). This is the continuous map of all the covered cycles in the program. The ACRIM Gap can be seen as a straight line (named on the graph) between 1979 and 1981.

From these findings, the ACRIM scientists performed what they considered to be the exact adjustments necessary to the findings, resulting in a uniform average waveform being established for the entire period, much along the lines just suggested (Fig. 5.4). However, the readings from the ailing NIMBUS-7 satellite required adjustment to offset the increasing error appearing in its record. Note in Fig. 5.5 that the red dots at the bottom of each

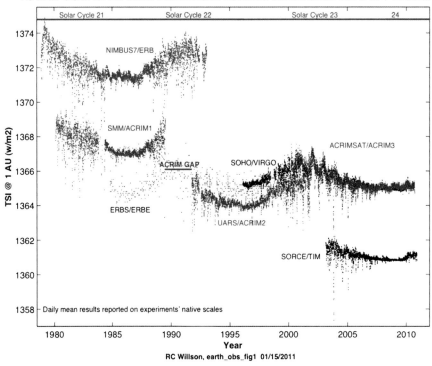

FIG. 5.4 The complete satellite record from 1980 to 2011 (Graph courtesy of NASA/ACRIM)

cycle show a lower point for Cycle 23 than Cycle 22. The upshot is that the period when warming trends were significant in the 1990s coincided with *increased* solar irradiance, according to the ACRIM researchers.

Next, compare the results of the adjustment made by the PMOD team in Switzerland, subtly different but quite the *opposite* to that of the ACRIM team. Their claim, thus, was that significant warming occurred during a time of *decreased* solar irradiance! Note how just a small difference of measurement produces a dramatically different result. This led to the PMOD team's entirely different conclusions about the larger issue of solar output (Fig. 5.6).

Although his theories of cosmic rays, and our journey through the galactic arms, led him to a new view of the ice ages, the leading

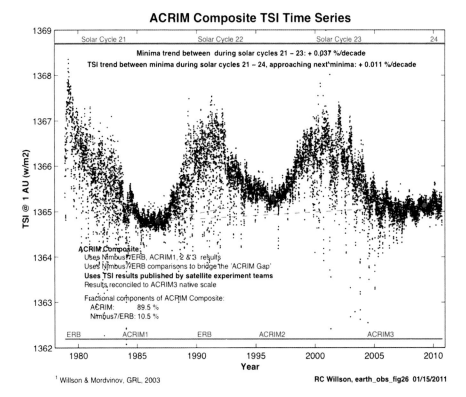

FIG. 5.5 The findings of the ACRIM team 1980–2011, with a continuous slope gradient of solar irradiance up to 2011. Note also the first indication of decrease during the present century, still higher than the first 11-year solar cycle on the graph but lower than the second. This coincides with many solar scientists' claims that the Sun has now entered a cooling trend since Cycle 23 (Graph courtesy of ACRIM/NASA)

researcher, Nir Shaviv, a long-time critic of the IPCC, was an unlikely supporter of the PMOD findings, illustrating disagreement even among those who oppose these views! [19]

Thus, with some still claiming that the PMOD interpretation is more likely to be accurate, the controversy still remains at the forefront of climate science to this day, and one of the most hotly debated topics. Remarkably, members of the public seem completely unaware of it. For those who are aware, the perception of a "gap of credibility" between researchers of different stripes must only have reinforced the uncertainties of climate science in general. It also reveals the degree of acrimony that has entered the dialog between colleagues, regretfully, becoming personal for

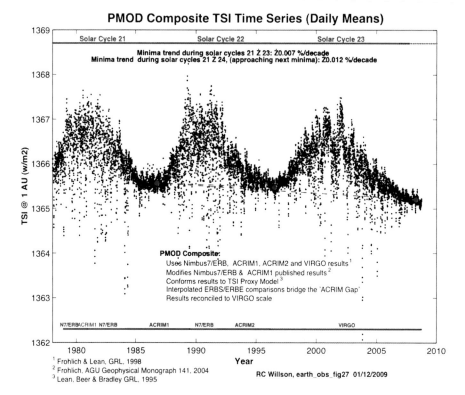

FIG. 5.6 The solar irradiance gradient according to the PMOD researchers (Graph courtesy of ACRIM/NASA)

some. For the rest of us, we have to build our own sense of where reality may really lie.

Recent NASA Climate Data

We cannot leave this part of the discussion without looking at some overall information from NASA Sun/climate research of recent years. Some have stated that there is little correlation between climate and solar activity, or that trends indicative of recent climate warming have not been detected in the atmosphere. The latter, at least, does seem contrary to NASA's graphs that show temperature change at various altitudes in the atmosphere since 1980 (Fig. 5.7). (Refer again to Chap. 2.)

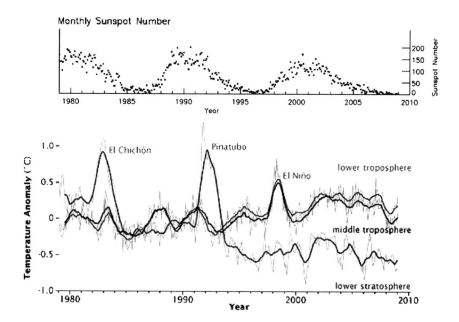

FIG. 5.7 NASA records from 1979 to 2010 (Graphs courtesy of NASA)

An international panel of experts led by NOAA and sponsored by NASA has released a new prediction for the next solar cycle. Solar Cycle 24 will peak, they say, in May 2013 with a below-average number of sunspots.

"If our prediction is correct, Solar Cycle 24 will have a peak sunspot number of 90, the lowest of any cycle since 1928 when Solar Cycle 16 peaked at 78," says panel chairman Doug Biesecker of the NOAA Space Weather Prediction Center.

We can see that the upper layers (stratosphere – shown in red) have cooled significantly, which should indicate a warming trend in the lower layers (troposphere – yellow and blue). Indeed, the graph does show this. Accounting for the infamous El Nino spike of the 1990s and solar peak in 2001, as well as the effects of volcanic aerosols, an overall warming trend seems clear. The temperature readings of the upper and lower atmosphere expected with greenhouse gases do indeed seem right in step with conventional warming theories. Whether the result here is due to increased CO_2, water vapor, other greenhouse forcing or feedback factors, or yet unexplained anomalies is, however, hard to determine, but thought provoking nonetheless.

Just before 2010, according to this graph, the troposphere did seem to have warmed to a greater degree than the stratosphere cooled, and before that the stratosphere cooled to a slightly lesser degree overall than did the troposphere. Thus, conclusions are not clear-cut. Readers must, of course, decide for themselves what all of this means, but it seems hard to doubt that the lower atmosphere is warmer today and the upper atmosphere cooler than it was in 1980.

References

1. Kobashi T, Severinghaus JP, Barnola J-M, Kawamura K, Catrer T, Nakaegawa T (2009) Persistent multi-decadal Greenland temperature fluctuation through the last millennium. Springer, New York; Masse G, Belt S, Sicre M-A (2010) Iceland in the Central Northern Atlantic: hotspot, sea currents and climate change, IUEM Plouzané, France, Arctic Ice: high resolution reconstruction; Bluemle JP, Sabel JM, Karlen W (1999) Rate and magnitude of past global climate changes. Environ Geosci 6(2):63; Keigwin LD (1996) The little ice age and medieval warm period in the Sargasso Sea. Science 274(5292):1504–1508

2. Hughes MK, Diaz HF (1994) Was there a medieval warm period, and of so, where and when? Clim Change 26(2–3):109–142; Jones PD, Briffa KR, Barnett TP, Tett SFB (1998) Millennial hemispheric temperature reconstructions, IGBP Pages/World Data center for Paleoclimatology; Mann ME, Bradley RS, Hughes MK (1998) Global-scale temperature patterns and climate forcing over the past six centuries. Nature 392(6678):779

3. Lamb HH (1965) The early medieval warm epoch and its sequel. Palaeogeogr Palaeoclimatol Palaeoecol 392, 1998 and 1999 extension to A.D. 1000. Meteorological Office, Bracknell, Berkshire, UK

4. Perry CA, Hsu KJ (2000) Geophysical, archeological, and historical evidence support a solar-induced model for climate change. Proc Natl Acad Sci USA 97(23):12433–12438

5. Budyko MI (1968) The effect of solar radiation variations on the climate of the Earth. Main Geophysical Observatory, Leningrad

6. Mann ME (2002) Little ice age. In: MacCracken MC, Perry JS (eds) Encyclopedia global environmental change. Wiley, Chichester; Mann ME (2002) Medieval climate optimum. In: MacCracken MC, Perry JS (eds) Encyclopedia of global environmental change. Wiley, Chichester

7. Moberg A, Sonechkin DM, Holmgren K, Datsenko NM, Karlén W (2005) Highly variable Northern Hemisphere temperatures reconstructed from low- and high-resolution proxy data. Nature 433(7026):613–617; Lean J, Beer J, Bradley R (1995) Reconstruction of solar irradiance since 1610: implications for climate change. Geophys Res Lett 22(23):3195–3198

8. Mann ME, Bradley RS, Hughes MK (1998) Global-scale temperature patterns and climate forcing over the past six centuries. Nature 392(6678):779

9. Solanki SK, Fligge M (1999) A reconstruction of total solar irradiance since 1700. Geophys Res Lett 26(16):2465

10. Soon W, Baliunas S, Idso CD, Idso SB, Legates DR (2003) Reconstructing climatic and environmental changes of the past 1,000 years: a reappraisal. Energy Environ 14:233–296

11. Cook ER, Palmer JG, D'Arrigo RD (2002) Evidence for a 'medieval warm period' in a 1,100 year tree-ring reconstruction of past austral summer temperatures in New Zealand. Geophys Res Lett 29

12. Moberg A, Sonechkin DM, Holmgren K, Datsenko NM, Karlén W (2005) Highly variable Northern Hemisphere temperatures reconstructed from low- and high-resolution proxy data. Nature 433(7026):613–617

13. Budyko MI (1968) The effect of solar radiation variations on the climate of the Earth,' Main Geophysical Observatory, Leningrad; Solanki SK, Fligge M (1999) A reconstruction of total solar irradiance since 1700. Geophys Res Lett 26(16):2465; Perry CA, Hsu KJ (2000) Geophysical, archeological, and historical evidence support a solar-induced model for climate change. Proc Natl Acad Sci USA 97:12433–12438

14. Shapiro AI, Schmuntz W, Rozanov E, Schoell M, Haberreiter M, Shapiro AV, Nyeky S (2011) A new approach to long-term reconstruction of the solar irradiance leads to large historical solar forcing. Astron Astrophys 529:A67

15. Archibald DC (2006) Solar cycles 24 and 25 and predicted climate response. Energy Environ 17(1):29–36 (Perth, Australia)

16. Adams P, Pierce J (2009) Can Cosmic rays affect cloud condensation nuclei by altering new particle formation rates? Geophys Res Lett 36(9)

17. Lockwood M, Fröhlich C (2008) Recent oppositely directed trend in solar climate forcings and the global mean surface air temperature. II. Different reconstructions of the total solar irradiance variation and dependence on response time scale. Proc R Soc A Math Phys Eng Sci 464(2094):1367–1385

18. Willson RC, Mordinov AV (2003) Secular total solar irradiance trend during solar cycles 21–23. Geophys Res Lett 30(5)

19. Shaviv NJ (2006) Carbon dioxide or solar forcing? ScienceBits. www. sciencebits.com/CO2orSolar

6. Gravitational Interactions of the Solar System

We now have examined some elements of the Sun's output and behavior, as well as its variability. These are merely a taste of the immensely complex and tangled interactions of what makes up the mechanics of entire Solar System. More significant are possibilities some researchers have raised that this mighty machine may directly and indirectly affect the Sun's output and Earth's climate in ways that are usually not considered, or presently possible to factor into projections (Fig. 6.1).

Do the Sun's Travels in Space Affect Its Output?

Since the Sun is not fixed in space, it is affected, to a small but not insignificant degree, by the orbital masses of the planets (and especially Jupiter and Saturn). These giant planets cause the Sun to orbit around what is termed its *barycentric center* (Fig. 6.2), in a motion forming an epitrochoid. An epitrochoid is a strange geometric animal; although superficially mimicking a circle (a closed system), it is actually an endlessly open-ended series of smaller loops subscribed inside larger loops. Ultimately it is a breed apart.

An epitrochoid also comes with its own unique parade of visual stunts, and to an observer in a fixed location it will cause the effect of retrograde motion from time to time. Imagine compass points north, south, east and west on the graphic (Fig. 6.2). In this instance the inner loops slowly work their way around the circumference until they approach the opposite side relative to our vantage point. If we are situated due southwest (where the Sun is pictured on the graphic), looking at the Sun when it is due northeast, the action of the inner looping will cause us to see the Sun's motion

A. Cooke, *Astronomy and the Climate Crisis*,
Astronomers' Universe, DOI 10.1007/978-1-4614-4608-8_6,
© Springer Science+Business Media New York 2012

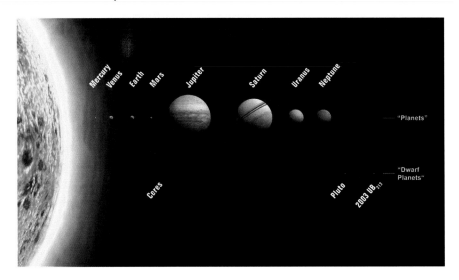

FIG. **6.1** The Sun's domain: keys to climate? (Graphic image courtesy of NASA)

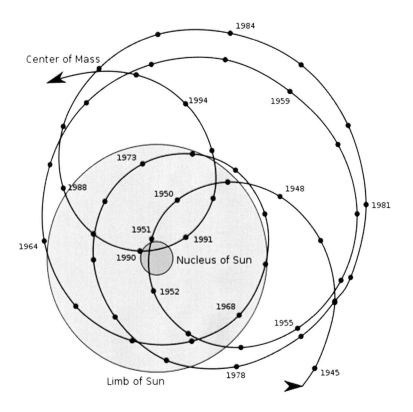

FIG. **6.2** Motion of the Sun around the barycenter of the Solar System (Graphic by Carl Smith)

as moving backwards relative to its constant clockwise movement around the barycenter.

The solar orbital path (over two solar diameters) may seem small relative to the Sun itself, but when we realize its diameter is over 1,392,000 km, this is not an inconsiderable distance for it to move in relation to the sizes and distances of the planets. Although a link between solar activity and its orbital mechanics has yet to be proven beyond a doubt, given that the Sun's core is far denser than its outer layers, the gravitational pull from the giant planets should cause the core and fluid-like interior to gyrate within the whole. Thus the varying distribution of the Sun's mass as well as its symmetry is directly affected, which, in turn, also affects planetary orbits. It seems only reasonable that it could also affect the inner processes of the Sun itself, and even perhaps produce measurable effects we might experience in Earth's climate.

Look at Fig. 6.2 again. Could it be more than a coincidence that approximately 11 years elapses between outer points of any imaginary complete double cycle of outer and inner loops, and the reversing polarities of each 11-year cycle? The concept behind such thoughts is not too far out of line with some ongoing research. Some scientists have directly tied the familiar sunspots with planetary interactions and the resulting internal solar upheaval, their manifestations becoming most prominent at those times of maximum tidal stress. If the solar cycle can be thus tied to the planets, we would expect to see a similar reflection in the movements of the Sun as they influence it.

In the 1980s, the pioneer Rhodes Fairbridge speculated on small changes to Earth's orbit of up to 1% and its impact on the climate [1]. According to his scenario, projected differences in distance to the Sun are the result of its barycentric orbit. However, common to all of these studies, one of the potential results seems eerily similar: significantly hotter or colder conditions, the orbital changes being perhaps one of the mechanisms triggering ice ages. As an early arrival (in the 1960s) to the study of the direct and variable influences of the Sun on the Earth/Moon system through gravitational and electromagnetic energies, Fairbridge was a pioneer. His work heralded investigations that would become central to the research of many others in looking for celestial links to climate.

In 2003, D. Juckett theorized that all of the Sun's activity could be explained by the link of its barycentric motion to the

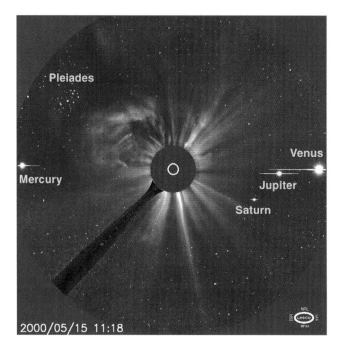

Fɪɢ. 6.3 The Sun and nearby planets (Image courtesy of NASA)

mechanics of angular momentum of the entire Solar System and its spin-orbit coupling [2]. We already know that even the domi-nant body also has its own orbital mechanics – as barycentric motion. Thus, for every given orbiting mass around another, there are appropriate orbital *and* axial turning speeds for both. It has been proposed that times of imperfect balance between the planets and Sun could be the trigger for solar activity as the Sun compen-sates for it by changing its rotational speed.

Spin-Orbit Coupling

This is the interactive gravitational process that controls orbiting and rotating bodies in space in achieving a balance of orbital and spin momentum. For any given distance and mass of such orbital bodies, there will be a balance between them whereby their individual orbital speeds and rotations are locked into a total synchronicity reflecting those factors.

In 2004, Habibullo I. Abdussamatov also trod increasingly familiar ground regarding potential external influences upon climate, thus speaking volumes about the scientific potential of new areas of research [3]. However, the reasons for greater attention not being paid to the concept of influences from beyond our own world in general are less clear.

The planets probably having the greatest gravitational influence on the Sun (and vice versa) are Mercury, Venus, Earth, Jupiter and Saturn, simply because of their proximity and/or mass. Figure 6.3 shows all of these planets but Earth, imaged in space by NASA's Heliospheric Observatory. The far-reaching energy of the Sun (flares and coronal mass ejections) is apparent despite the blocking of its disc; it still dominates the view, along with activity in its corona and beyond.

Solar Variations

We can project from Fig. 6.2 approximately where the Sun may presently be located in relation to previous years. According to the chart, it has just entered a phase of greatly reduced output and is considered to be in the second solar orbit within the larger nine-orbit solar cycle. At this time early in Cycle 24, there has been a noticeable decrease in sunspots, solar wind, and in radiant output in the amount of 0.02 % in the visible portion of its spectrum, and about 300 times as much in the extreme ultraviolet portion. (See again the discussion in Chap. 4 referencing disproportionate increases in UV during active years).

During cold periods solar output reductions may seem small in relation to the total. However, we have already seen that some researchers believe that effects of this level of change on Earth's climate may be more significant than they might appear. They have theorized the possible lowering of temperatures *indirectly* by yet to be defined mechanisms, and not necessarily by just the reduced influence of the disproportionately reduced high frequency UV radiation. Additionally, a downside of a less active Sun is a greatly reduced solar magnetic heliosphere, generally considered to be a protection against cosmic radiation reaching Earth. Cosmic rays have been implicated in cloud formation by other research, and the possible mechanics of this process will be discussed in Chap. 12. As would be expected, this related sub-topic has generated its own field of controversy.

A Cooling Period?

In definitive positions, such as can be found on the website skepticalscience.com it is stated that there is no scientific basis to claims that the planet (Earth) will enter a cooling phase in the near term. Whether or not Earth is actually entering such a phase, it can be shown that there is indeed a valid scientific basis at the very least behind proposing such a scenario.

A Delicate Balance

Meanwhile, other researchers have continued to explore the possibilities of an Earth/astronomical connection. In a presentation at the 2007 American Geophysical Union fall meeting, Ingo H. Leubner demonstrated a quantitative model to reflect the degree that planetary orbits can be affected by even slight changes to the Sun's radius [4]. Leubner spelled out some grisly specifics. Because the Sun has been shown to vary by up to 1% in radius, his model showed a directly linear, and remarkably sensitive, correlation with radically altered planetary orbits.

Planetary Orbital Mechanics

A curious eighteenth-century theory named Bode's Law proposed that the orbital radii of the planets in the Solar System should correspond to a precise formula related to the standard astronomical unit (AU). Discovery of a new planet was considered imminent at the distance from the Sun of 2.8 AU, and the name Ceres was selected in advance. Although Bode's theory ultimately was rejected, it was not before one of the largest objects in the Asteroid Belt had been wrongly assigned as the discovery and given the preselected name.

His scenario for Earth was anything but rosy. With increased solar radii corresponding to wider orbits and cooler climates – and vice versa – amazingly small increases and decreases of the solar radius were shown for Earth to move potentially close enough to the Sun to destroy all life, or at the other extreme (of only 0.47% solar radius increase) to release it from Solar System orbit (also destroying all life)! Leubner also linked lesser variations to past ice ages, but it was interesting that he did not dismiss AGW theories, instead seeing their value in combination with others.

Leubner estimated margins for such disasters. They are far less than most people might ever realize. Per 0.001% solar radius change, a temperature variance of 0.29°C, plus 0.90°C per 0.67% of orbital change could be expected – in either direction – depending upon whether the change is positive or negative. It would also add or subtract more than a day from the calendar year. Furthermore, a 10% solar radius decrease would be sufficient to bring Earth almost to the Sun's surface! That serious and respected figures have actually proposed linking to climate anything remotely connected to external influences from deep space may come as a surprise to those who have only contemplated the human role in the equation. Even more clear is that the most microscopic change in the Sun's radius could affect the length of day (LOD) by as little as a few seconds or minutes.

The 179-Year Cycle

Although the chaotic (the exact opposite of linear) nature of the planets' own orbital motions ensures that the path the Sun takes will never quite duplicate itself, an overall period of about 179 years elapses before all the planets are in approximately the same positions again. In this time, the Sun will have completed about 9 epitrochoidal circuits around the Solar System's barycenter. Also, once during this period the planets will have all aligned on one side of the Sun, producing a tidal pull of potential significance. Ching-Cheh Hung, of the NASA Glenn Research Center and whose research we will return to in greater depth in Chap. 7, attempted in 2007, with some success, to demonstrate that the tidal effects of the planets are likely primary forces behind solar activity [5].

Some believe it is possible to identify four such 179-year cycles in the last eight centuries, when there have been periods of cold on Earth at their minima. It has been speculated that these historical periods may parallel the beginning of the present cycle (including the Dalton and Maunder Minimums) and that similar climatic consequences might be in store.

Curiously, Burroughs (see Chap. 1) pointed out that there is another recognizable period of tidal resonance between Earth and the Moon of almost exactly the same amount of time as this 179-year period at high latitudes. Whether this is more than a coincidence, Burroughs also identified the 179-year solar alignment cycle (along with many more), but seemed to miss a potential connection with the almost identical period of the 179-year lunar tide and perhaps some further possible connection to that of the Sun. Stating that the Moon has a more profound influence on Earth than other external cyclical influences (and therefore possibly on its climate), if this was an oversight it seems surprising, although the 179-year number may be just a coincidence he was well aware of.

However, Burroughs was not the first to speculate on that link. As early as 1965, Paul. D. Jose addressed the 179-year Solar System cycle, in addition to its interaction of the 11-year sunspot cycle [6]. In 1974, another research paper, by Cohen and Lintz, examined the larger cycle and found that it has a regular frequency that might be seen in the power spectra of solar activity [7]. However, over the years all of this early groundwork research regrettably seemed to disappear from view.

The 60-Year Cycle

Regardless of one's sentiments about the causes of climate change in general, there is more than enough reason to look further at what has been loosely termed the 60-year cycle. Although the precise number of years regarding its length is more of an average than a precise figure, one can continue to find almost endless references to the pattern in scientific literature.

If the 60-year cycle has continued to occupy attention amongst those looking for larger connections to the sun, it is because it appears to be the most prominently shown among temperature

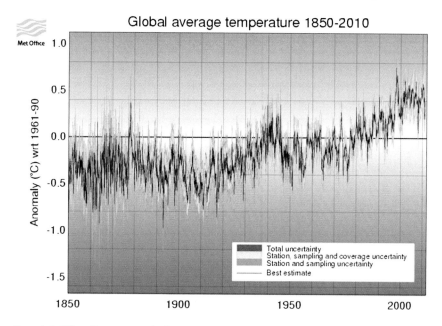

Fig. 6.4 The 60-year cycle (Image courtesy @ British Crown copyright 2010, the Met Office)

and proxy records, and thus is the most potentially significant. Some believe that, due to the current theorized waning of the cycle, this one external driver alone could largely counteract any warming trends over the next few decades, and that temperatures might be expected to increase by the end of the century by not very much at all. Compared to the temperature increases indicated by most climate models, this is in line with the projections obtained by entirely different means (!) by T. J. Nelson, referenced in Chap. 2. These more modest climate scenarios exist in stark contrast to many scarier versions projected in most conventional AGW theories.

We can detect the cycle when examining many graphics, even when they are not specifically constructed to show the phenomenon. One example is the temperature graph (Fig. 6.4) from the Hadley Climate Research Unit in England, showing historical readings from 1850 to the present. Although its purpose was merely to show the record, it has been used frequently in articles and websites to illustrate the possible existence of the cycle.

There are limitations to what we can infer, of course, imposed on us by the relatively short time scale of accurate record keeping.

However, the 60-year cycle does seem clearly evident on the Hadley chart (approximately 1880–1940 and 1940–2000) and contained within an overall upward temperature trend. It is not always so clearly revealed on many other graphs. The El Nino 'spike' of 1997/1998 is quite clear and draws our attention to similar 'spikes' at near comparable times in previous 60-year patterns. Historically, a strong El Nino event was registered in 1946, right in line with the highest such 'spike.' Similarly, as if right on cue with the previous high mark, there was a notable increase of El Nino events in the late 1870s, corresponding with a period of extreme drought and famine in Asia at the time.

If we accept the possibility of a 60-year or so cycle, it would appear that in 2011 we are occupying the downside of the last such cycle, and *might* thus be seeing possibly the first clear stages of a decline in warming. The full down and up cycle will bring us to about 2060. Although the overall approximate 1°C climate warming throughout most of the preceding twentieth century is not an item of controversy, we can see that this is also clearly shown on the graph by its continual upward trend.

Researchers and their theories concerning the 60-year cycle are numerous, but among those who have inferred its existence include the following:

- Timo Niroma made explicit mention of such links in publications on his personal website [8]. He paralleled the larger historical periods of sunspot activity (or lack thereof) that coincide with the 60-year cycles of temperature. As he saw it, the 60-year timings are dead on, although they do not necessarily alternate in sequence. There often can be more than one cold or warming cycle in a row. He even listed the dates through the past 400 years that correspond to solar cycles; certainly these do seem to approximate the known record.
- Craig Loehle, in his 2009 paper about the solar satellite record, referenced many studies; 50- to 70-year cycles featured prominently in his discussions and analyses [9]. Loehle concluded and demonstrated that once the warming from cycles is removed from trends shown in climate models, the upward curve bears no resemblance to the warming and cooling periods under examination.

- Biondi et al. (2001) were able to tie proxy data to the historic Pacific Decadal Oscillations and the El Nino Southern Oscillation, not only confirming historical periods of climate change but reconciling these findings with an approximate 50- to 70-year cycle [10].
- Klyashtorin and Lyubushin (2003) went even further [11]. They first showed 60-year cycles, consisting of alternate periods of warming for 30 years, followed by 30 years of cooling. In relating these statistics to the consumption of fossil fuels during each 30-year span, they found little correlation, while noting an overall warming trend from 1861. Utilizing data from ice cores and their power spectra, they calculated it was possible to accurately deduce a 64.13-year cycle from it. By this they were able to produce a model of sorts that implied cooling from 2005 for the next 32 years (half of the cycle). We may soon see – over the next decade or two – if their estimations resemble reality in any way.
- Michael E. Schlesinger and Navin Ramankutty similarly recognized this cycle, but as a 65- to 70-year period, maintaining that a broader cycle in the northern Atlantic Ocean of 50–88 years is responsible [12]. By applying what they term as 'singular spectrum analysis' they believed that this phenomenon has made the degree and effect of any anthropogenic warming difficult to measure.
- Nevertheless, in a follow up 2000 paper, Schlesinger and his co-author concluded, seemingly in contradiction to the 1994 position, that despite these oscillations, anthropogenic greenhouse gases are the primary mechanism behind present-day warming trends [13].
- In another interesting parallel, another study showed that Earth also reflects a 60-year cycle in its small length of day (LOD) fluctuations [14].

Taking the Hadley graph and others, some interested parties have overlaid successive 60-year cycles upon one another, eliminating the warming trend in order to represent the same base level temperature, and found remarkable similarities between the cycles. However, decisive conclusions remain elusive. Ongoing disagreements continue about the accuracies of temperature records on

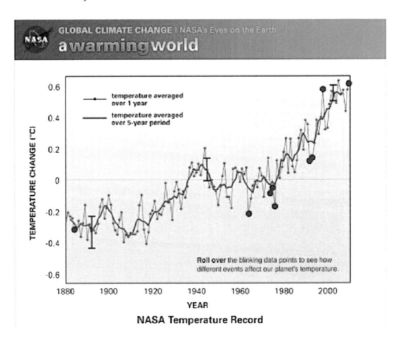

FIG. 6.5 Global temperatures (Courtesy of NASA GISS)

which these graphs are based, the reliability of contemporary terrestrial measurements versus satellite readings, the 'urban heat
island effect' (more on this in Chap. 8), different warming trends in
the oceans, influences on varying cloud cover, the reality of the
cycle itself, and so on. Such is the hard-to-define nature of climate
studies.

Let us look further to records from the Goddard Institute
(GISS), with the global temperature graph from a 2001 study [15]
covering essentially the same period (although stopping at year
2000), but updated to reflect their revised methodology (Fig. 6.5). It
should be pointed out that the official position of the Goddard
Institute on climate change is closely aligned with that of the
IPCC. The resulting graph does not draw attention to the supposed
60-year cycle that shows so well on the Hadley chart, and we come
away with a rather different impression. Indeed, taking this graph
in isolation we would probably be unaware of any cyclical
elements at all. And the ongoing rising pattern without cyclical
elements is more strongly accentuated on the GISS graph by the
selection of a taller vertical scale.

However, it is worth noting that although the IPCC has yet to acknowledge any possible indirect warming effects of the Sun, similarly they are yet to acknowledge most of the repetitive cycles that many researchers believe show fairly clearly on the record. At least they have recognized the 11-year sunspot cycle, and in the 2007 report (AR4) they have now included another possible cycle at 100 years, and still another at 140 years in their analyses.

Thus, there seems to be some late recognition that cycles may be evident in climate; even so, these long-term cycles have not been significant in their assessments about present climate change issues. It is these very phenomena that some contend could be important keys to understanding climate on Earth. But there are other considerations in making conclusions from any particular record. On the previous graph (Fig. 6.5), the first peak has already occurred before the timeline begins; the second at 1940 is certainly evident. But that of the late 1990s is less so; indeed it appears that the peak has not yet been reached. However, look closely. The high spike of 1998 is there, but there is no record much beyond that, other than a couple of years of temperatures similar to the approach to 1998.

Taken in isolation, the GISS graph does not suggest the 60-year cycle, unless we are looking for it, having seen the Hadley chart. The temperature peaks around 2000 and suggests an overall ongoing and radical ascent into the present time, one that seems to be on a relentless upswing, albeit with a few bumps along the way. The impression is thus quite different, largely because of the absence of the key starting and ending dates shown on the Hadley graph. It is worth noting, too, that GISS researchers have continually modified their methods over the years in ways that increasingly seem only to reflect their evolving model! Increases culminating around 1940 have been gradually lowered, while more recent temperatures have been increased. This has not played well with their critics, despite the justifications given.

In another NASA graph (Fig. 6.6), taken from four records from around the world (including the Hadley Climate Research Unit), the 60-year cycle perhaps may be more evident. Of course, this may only be so to those who accept the existence of the cycle, while not necessarily being convincing to everyone else. However, in separating the various records shown on the graph,

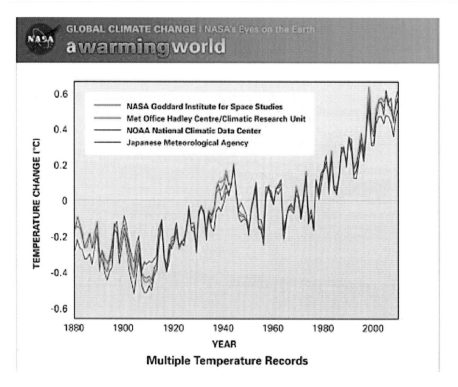

FiG. 6.6 Multiple temperature records (Courtesy of NASA GISS)

compare that of the Japanese Meteorological Society and the Met
Office Hadley Center with that of NASA Goddard. That record
seems more closely aligned to the original Hadley graph of Fig. 6.4
than do the others, in regard to the low point around 1910 and
the high point after 1998. Both of these records make the high
point of 1949 register more in accord with the outline of the
60-year cycle.

Should the 60-year trend be a reality, the peak of the most
recent cycle ought to have occurred by now. However, some mixed
readings have made that difficult to determine, together with the
need for more time to elapse for it to become clear in its descent.
If the event has indeed peaked and does show up clearly over the
next few years in readings, it would certainly be evidence of the
60-year cycle being a component in present climate conditions. In
2009, Craig Loehle discussed satellite readings of the period from
the late 90s until 2009 that showed cooling trends during that time
[9]. A modeled projection by Klyashtorin and Lyubushin that
appeared in their 2003 paper and extending to 2009 does indeed

align quite accurately to observed satellite data [11]. This, of course, still is too narrow a period to determine anything with certainty.

If a cooling trend during the second decade of the twenty-first century is not born out over time, some researchers have suggested that Long-Term Persistence (LTP) may be responsible. Loehle did point this out in comments about the study by Easterling and Wehner [16]. Here, these authors postulated that should a cooling trend be occurring during a time when greenhouse warming is set in a pattern, perhaps LTP might take place. This would keep the temperatures in a basic warming mode that could effectively mask the cooling trend. It was stated that any such trend might be incidental to natural cycles (at least this is an acknowledgement of such phenomena) and not necessarily a refutation of AGW. Many would argue that if the climate fails to cool, those who use the LTP argument are merely finding their way out of an ideological jam. However, Easterling, who proposed it, and is a colleague of Hansen, is firmly in the camp of those who support the theories of AGW!

Clearly, although it has been problematic to establish to everyone's satisfaction that AGW really has taken hold, it seems clear that Loehle is among those unconvinced of its validity. With his remarks referring to cycles approximating the 60-year cycle, he concluded his paper with the observation that the largely *non-linear* warming observed since 1880 is out of step with the much more uniform rise in CO_2 levels. This last point certainly can be seen from the records, regardless of any explanation. It is also central to the case against existing climate models and AGW. In challenging these models, Loehle stated it is likely that we have now entered a several decade-long period of cooling – or at least of stable temperatures.

Loehle also pointed to satellite temperature records that show a cooling trend starting during the first decade of the new twenty-first century. It is significant that satellite records differ in many ways to other means of measuring temperature. Thus, again he brought the accuracy of existing climate models into question, which have varying predictions of ongoing substantial increases in temperature. Perhaps an even more provocative position was his view that what had been experienced as more rapid warming during the last few decades of the twentieth century could have been caused by a combination of external cycle forcings. These would have exaggerated any possible influence of AGW.

However, perhaps the single most fascinating attributes common to so many research papers and articles, not all necessarily to do with climate, are the repeated references to a recurring 60-year cycle. Craig Idso and Fred Singer, in their enormous 868-page report for the counter-IPCC group, considered that the Sun's influence consists of complex feedbacks not yet understood but real nonetheless [17]. Using the Hadley temperature graph shown in Fig. 6.4, they superimposed the projections of the IPCC climate model of 2007 (see Chap. 8). Only recent projections of warming seemed to correspond with actual observed temperatures, and Idso and Singer commented that the 60-year cycle was missing.

> From the homepage of the website of the counter-IPCC group:
> The Nongovernmental International Panel on Climate Change (NIPCC) is what its name suggests: an international panel of nongovernment scientists and scholars who have come together to understand the causes and consequences of climate change. Because we are not predisposed to believe climate change is caused by human greenhouse gas emissions, we are able to look at evidence the Intergovernmental Panel on Climate Change (IPCC) ignores.

The 60-year periodicity shows in many additional guises: an assortment of records shows it in meteorites in China [18], or in beryllium and carbon isotopes [19], among others [19]. In these papers, all of the proxy records are compelling and indicative of such a cycle, though perhaps not conclusive to every researcher as being a controlling influence on climate.

Perhaps most notable among other researchers' findings, although not necessarily related to the 60-year cycle, per se, is that of Ching-Cheh Hung, in which he drew direct links with the effects on the Sun by some of the planets [5]. We will look closer at that special corner of astronomical research in Chap. 7.

References

1. Mackey R (2007) Rhodes Fairbridge and the idea that the sun regulates the earth's climate. J Coast Res (Special Issue) SISO 955–968
2. Juckett D (2003) Temporal variations of low-order spherical harmonic representations of sunspot group patterns: evidence of solar spin-orbit coupling. Astron Astrophys 399(2):731–741

3. Abdussamatov HI (2004) About the long-term coordinated variations of the activity, radius, total irradiance of the Sun and the earth's climate. International Astronomical Union (IAUS) 223:541–542
4. Leubner IH (2007) Variability of stellar and solar radii and effect on planetary orbits and temperatures. American Geophysical Union Fall Meeting 2007
5. Ching-Cheh Hung (2007) Apparent relations between solar activity and solar tides caused by the planets. Report NASA/TM-2007-214817. Glenn Research Center, Cleveland
6. Jose PD (1965) Sun's motion and sunspots. Astron J 70:193–200
7. Cohen T, Lintz P (1974) Long term periodicities in the sunspot cycle. Nature 250(5465):398–399
8. Niroma T. The effects of solar variability on climate. 2001. http://www.tilmari.pp.fi/tilmari5.htm
9. Loehle C (2009) Trend analysis of satellite global temperature data. Energy Environ 20(7):1087–1098
10. Biondi F, Gershunov A, Cayan DR (2001) North Pacific decadal climate variability since 1661. J Climate 14(1):5–10
11. Klyashtorin LB, Lyubushkin AA (2003) On the coherence between dynamics of the world fuel consumption and global temperature anomaly. Energy Environ 14(6):773–782
12. Schlesinger ME, Ramankutty N (1994) An oscillation in the global climate system of period 65–70 years. Lett Nat 367:723–726
13. Andronova NG, Schlesinger ME (2000) Causes of global temperature changes during the 19th and 20th centuries. Geophys Res Lett 27:2137–2140
14. Yu Z, Roberts P, Russell CT (2004) Correlated changes in the length of a day and the magnetic field with a period of 60 years. American Geophysics Union, Fall Meeting
15. Hansen J, Reudy R, Sato M, Imhoff M, Lawrence W, Easterling D, Peterson T, Karl T (2001) A closer look at United States and global surface temperature change. J Geophys Res 106:23947–23963
16. Easterling DR, Wehner MF (2009) Is the climate warming or cooling? Geophys Res Lett 36:L08706
17. Idso C, Singer SF (2009) Climate change reconsidered. The report of the nongovernmental international panel on climate change. Heartland Institute, Chicago
18. Yu Z, Chang S, Kumazawa M, Furumoto M, Yamamoto A (1983) Presence of periodicity in meteorite falls. Mem Natl Inst Polar Res 30 (Special issue):362–366
19. Ogurtsov MG, Nagovitsyn YA, Kocharov GE, Jungner H (2002) Long period cycles of the sun's activity recorded in direct solar data and proxies. Solar Phys 211:371

7. The Possible Effects of Solar Cycles

The search for hidden links to the Sun's activity has continued in spite of denials of that possibility by some. Although the potential for planetary interactions with the Sun has featured in some of these searches, to date, their effects on Earth's climate have been problematic to prove. However, in due fairness, it must be pointed out that many scientists do not consider that the IPCC position has been proven either. Even the IPCC itself refers to various levels of confidence in their positions rather than expressions of absolutes. The contention by Duffy et al. [1]. that twentieth-century warming can all be explained with established causes will not satisfy many critics of the IPCC stance either, since all of those known forcing influences are strikingly out of step with the rate of warming itself.

Other researchers also have suggested, indeed strongly implied, that the warming effects of the Sun more likely are produced indirectly rather than directly. If this sounds familiar, you may recall similar hypotheses from earlier in this book. Even as early as 2001, in his book Burroughs speculated about the possibilities of still undefined processes in the atmosphere amplifying small increases in solar irradiance into more significant warming effects. However, at that time, scientific consideration of any influences beyond just the 11-year solar cycle was still largely new and revolutionary.

Accusations of failure to include all factors such as these in climate models have been among the continuing objections to these projections ever since they first appeared. It has been pointed out repeatedly that modeling of changing climate conditions and temperatures over the past 40 years has only been possible after *assumed* contributing factors (i.e., anthropogenic additions) had

A. Cooke, *Astronomy and the Climate Crisis*,
Astronomers' Universe, DOI 10.1007/978-1-4614-4608-8_7,
© Springer Science+Business Media New York 2012

been accommodated into the models, and applied only from 1970 onwards. Critics claim the precise effect of anthropogenic greenhouse gases has therefore been deduced, rather than observed or proven through empirical data. Thus, they maintain that the underlying cause of the most recent temperature increases is far from established. We will continue this discussion in Chap. 8.

However, in the absence of definitive proof of conventional AGW, its proponents have only been able to go so far as to consider their own positions 'likely.' If proposed alternative theories are considered 'unlikely,' climate skeptics maintain that such a position does not translate into proof of it either, but only an opinion about as good as theirs. It is easy to see why controversies continue unabated. Seen from the outside, the stakes appear more in the realm of 'winner-take-all' than in solving the riddle. But first, everyone has to concede the existence of the riddle itself.

Now that the 11-year solar cycle has met with general acceptance, at least, we will look into those larger solar cycles discussed in the previous chapter. The gravitational interactions of any individual planet on the Sun would seem insufficient to cause such cycles. Besides, their orbits do not directly correspond to them, so we have to look further. Indeed, some have theorized that the cycles could result from the combined effects of the larger planets on the Sun, gravitational and/or otherwise. This has resulted in speculation and even theories that the 60-year cycle, in particular, could be linked to the Sun's extended relationship to climate, and that we should not only continue to investigate these phenomena thoroughly but probably include them in future projections of Earth's climate.

Thus, in addition to the 60-year solar cycle, other cycles involving the Sun include the 179-year cycle (see Chap. 6), the 70–100-year Gleissberg cycle (see Chap. 4), the 11-year cycle and 22-year Hale cycle (again, see Chap. 4), together with the possible existence and effects of still more; these all compound the situation. Even the 9.1-year lunar cycle is yet one more to consider. The exact mechanisms by which any of them might interact are not yet determined, let alone understood. Regardless, these cycles do not seem in evidence within most climate models, even absent entirely – an increasingly familiar-sounding comment. However, it is also fair to ask how they could possibly have been expected to be included, given the present state of knowledge.

The 9.1-year lunar cycle poses some interesting questions, too, although none yet seems to lead us directly to climate. Like the Sun and planets, the interactions of Earth and the Moon also result in a barycentric orbiting pattern between them, with subsequent variations of stresses. This produces direct affects beyond the obvious daily tides of the oceans. Sea levels also undergo substantial rises and falls, and as such, lunisolar tides can be seen in the phenomena of 'spring tides' that often result in coastal flooding.

It remains hard to know, however, what, if any, effect the lunar cycle might play in the larger climate picture. Certainly its effects appear strictly localized to the vicinity around Earth, but speculation has long existed of other possible lunar influences, from tectonic activity to contributing to the mixing of the deep ocean currents; we already know changes in the thermocline alter the weather on a regular basis in the Pacific, by altering warm and cold water distributions. Bearing in mind that we still do not fully understand what drives the oceans, could the Moon's interactions with them possibly have presently unseen effects and alter the climate, too? Again, at this time, such unknowns cannot be expected to be included in climate models.

Meanwhile, other forces are in play, too. Earth's fractured tectonic plate structure, the lithosphere – the outer shell that shifts – twists and rotates around Earth's two concentric inner cores. Outside forces, including the magnetic field of Earth and its interactions with the magnetic field of the Sun, even the Moon, might be linked to secondary effects on Earth, quite separately to those known consequences of tidal forces. Could it be that earthquakes might also follow this path? With the fragility of Earth's crust, the questions are endless, beyond climate itself, when one thinks it through (Fig. 7.1).

The Trigger Behind Sunspots and Solar Flares?

Many researchers long ago rejected the idea that large solar cycles could govern or influence the climate and chose long ago to exclude them from further consideration. Ching-Cheh Hung is among some noted scientists who have taken the opposite position, having stated that it is highly unfortunate that recent advances in solar

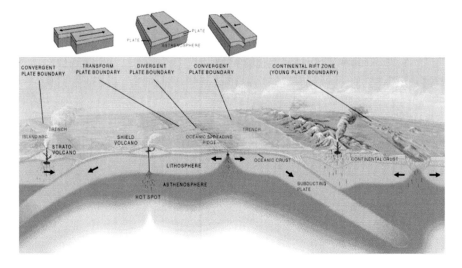

FIG. 7.1 Tectonic plate boundaries and activity along them (Courtesy of the U.S. geological survey)

output measurements (notably from now available satellite research such as the ACRIM program) have not been incorporated into climate modeling [2].

Hung approached his topic by attempting to understand specifically how the planets might interact with the Sun to create measurable variability of its output through flare activity. His detailed and complex findings seem significant indeed, opening with the observation that most of the largest known solar flares can be shown to have occurred under clearly observable circumstances. Almost predictably, it seemed, flares took place when one or more of the four most likely planets to have strong gravitational interactions were within a linearity of 10° of the flare – or, interestingly, on the *opposite* side of the Sun. Hung calculated the likelihood of the many solar flare events being merely coincidence to be near 0%. With the statement that the 11-year alignment of these planets correlates with the sunspot cycle, it was sure to attract attention (Fig. 7.2).

Since his main interest was the relationship of solar flares to disruptions of electronic systems on Earth and in space, researched through an effort to better protect our growing dependencies on

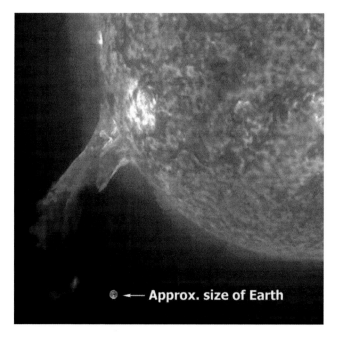

Fɪɢ. 7.2 Solar flare with relative size of Earth for comparison (Image courtesy NASA)

them, Ching-Cheh Hung did not bring a predetermination to prove or disprove any existing climate change theories. This in itself would seem to add considerable credibility to his findings. However, they form a link by default through their related areas of study.

Solar Flares

Solar flares are sudden fiery projections from the Sun's surface, representing enormous releases of energy of as much as a sixth of the total solar output each second! Plasma loops between spots are frequently precursors to flares, and are responsible for vast concentrations of electrons, ions and atoms flooding into space.

Further, Hung was able to implicate large variations in the solar magnetic field, extreme concentrations of which may be observed as sunspots. He referenced the connection between solar flare activity and solar variations previously studied in the twentieth century, expressing deep regret that they had largely been put aside since that time. In citing such studies going back to 1942, he indicated that his work was undertaken to advance the earlier research, with modern satellite monitoring becoming available after 1981. Accordingly, his unequivocal statement that a direct correlation can be made between solar flares and solar tides induced by the planets is highly significant. Such an uncompromising sentiment is strong stuff. By first illuminating the reasons why there are small variations in the combined tides of Mercury, Venus, Earth, Jupiter and that of the regularly occurring solar tide cycles, Hung was able to tie the 11-year and 22-year Hale cycles directly to them.

Unfortunately, Hung found that numerous other earlier studies had failed to note such relationships. This had led to existing conclusions that planetary tidal pull is too insignificant to have any measurable effect. Specifically, it was thought to be too regular, not closely enough aligned to the sunspot cycle, and too out of step with larger periods of cooling when the same planetary motions were also in play (such as the Maunder Minimum and others that occurred within the last 400 years). Hung believed the reason for that was due to some of the research being misinterpreted. By applying the wrong criteria, he believed that Jupiter's influence, in particular, largely had been missed by researchers.

By analyzing solar flares over 30 years and their relationship to known planetary positions, Hung was able to calculate the probabilities of predictability versus random events. His final conclusions about solar flares and sunspots, in essence, were:

- Almost all flares occurred when a planet was directly aligned within 10° of the Sun or on its opposite side. Accordingly, he considered the likelihood of flares occurring by pure chance not to be credible, and indicated that it should be possible to develop a system to predict solar activity.
- Flares typically occurred before alignment, although sometimes the greatest examples occurred when the planet in question was up to 30° from optimal position. In fact, the largest

known flare took place when Venus and Jupiter, the planets of greatest gravitational influence on the Sun, were both at such (30°) positions from the event plane, implying some process(es) not yet understood. This 30° figure, in regard to positioning of any of the four planets of most influence, cropped up several times with many of the largest flare explosions. Hung speculated that at these times, vertical and horizontal gravitational forces became significant enough to cause the events.

- In most instances, he noticed, the largest sunspots appeared to generate the greatest flares when they rotated into a place on the solar surface most directly affected by planetary position, whereas smaller flares did not seem to be necessarily influenced by tidal forces of the planets.

Interestingly, in 2009, the maverick researcher, Timo Niroma, made related but not exactly parallel comments regarding Jupiter's eccentric orbit, its magnetic/gravitational influence and effect on sunspot cycles [3]. Knowingly or otherwise, he seemed to concur more or less exactly (with Hung's conclusions) on the direct connection between them, maintaining at the closest point of approach that sunspot numbers are at a minimum, only to increase as a result of magnetic interactions before and after the approach. This seems to tie in with the 30° orbital placements reported in Hung's paper. Niroma's study, however, though again 'only' his personal webpage and not constituting peer-reviewed research, is remarkable in its complexity and depth of analysis, especially where it apparently coincides with the more recognized research of others. However, his independently produced theories are far from agreed upon, let alone accepted. They might not even be widely known, though they are worth at least a cursory inspection.

Meanwhile, Ching-Cheh Hung continued to explore the possibilities of planetary influences on the Sun occurring in cycles, where the combined effects of more than one exerts its greatest pull – as in the case of the 11-year solar cycles. Thus, perhaps multiple compounded cyclical gravitational effects were perhaps the key, and not those caused by the most obvious of them. Careful analysis, therefore was how Hung able to prove virtually a direct correlation with the combined cycles of Venus, Earth and Jupiter and the 11-year cycle. When the influence of Mercury was included, the clear correlation disappeared, so he determined that it did not affect the cycle.

Despite the near exact parallels between the sunspot cycles and those of these three planets, small irregularities remained. We can presume again that the absence of exact predictability (the 'chaotic' nature) of the Solar System is responsible. Hung was able to demonstrate that maximum sunspot activity seemed to occur with the most precise alignments and was thus able to explain the many periods of alternately low and high solar activity.

In that he further seemed resolute in asserting the likelihood that his determinations were correct, versus the opposite position that no such influences take place, in this respect, Hung's research would appear to be almost, if not entirely, unique, and his opinions worthy of note. On the strength of it, there seems sufficient reason to look closely at the possibilities beyond standard explanations of the Sun's influence in most climate models; it should have provided sufficient encouragement to recommend continued development of his thesis. Like many others, Hung was not alone in predicting that the present sunspot cycle (Cycle 24) will be very low in activity. Could it be that the possible, but still elusive, 60-year cycle is directly related to Hung's research?

From NASA:

An international panel of experts led by NOAA and sponsored by NASA has released a new prediction for the next solar cycle. Solar Cycle 24 will peak, they say, in May 2013 with a below-average number of sunspots.

"If our prediction is correct, Solar Cycle 24 will have a peak sunspot number of 90, the lowest of any cycle since 1928 when Solar Cycle 16 peaked at 78," says panel chairman Doug Biesecker of the NOAA Space Weather Prediction Center.

Hung left US with one more interesting prospect to mull over. Should the Sun's present low activity not be prolonged, the slight *lack* of synchronization of the three dominant planets would be canceled. Higher activity would be the likely result instead, through increased combined planetary interaction. Again, regretfully this allows for the potential to continue for many interpretations and positions.

Further Possible Interactions with Earth and Its Climate

At this stage, we might well be asking for more insights about how other orbital oscillations of the Solar System might interact with Earth's climate. Possibilities include potential forcing and feedback effects from the tidal influences of other planets on the Sun's sunspot activity and even output, including the effects of the solar magnetosphere, as well as its interaction with cosmic rays (see Chap. 12), the changing length of day (LOD – see again Leubner [4]), and even unknown or anticipated lunar tidal effects on the oceans, among many. Regardless, the fact that no link has yet been definitively proven does not mean varieties of larger combined influences could not exist. It is the prospect of these, perhaps in sum or working together to amplify their individual influences to produce a greater whole, that is behind these ideas.

Furthermore, the effects of what may seem to be minor influences might be determined also by their combinations, not only by the amount. If this is so, the total equation could be a subtle synchronicity between known forcings, feedbacks, together with yet to be determined and multiple celestial oscillations acting together in concert. There is an illuminating demonstration of the effect of synchronicity between otherwise separate entities – that of bringing the beat of multiple independently metronomes together – inspired by the demonstration of astronomer-mathematician Christiaan Huygens (1629–1695), of synchronizing two pendulum clocks on a common support. It is a simple experiment well known to physicists (Fig. 7.3).

We can readily perform this for ourselves or witness this effect among numerous examples on YouTube, for example:

http://www.youtube.com/watch?v=Aaxw4zbULMs

In this instance, five clockwork metronomes, set to the same number of beats per minute, are put in motion on a flat platform. Then the entire group is placed on a 'fluid' base – two cylindrical

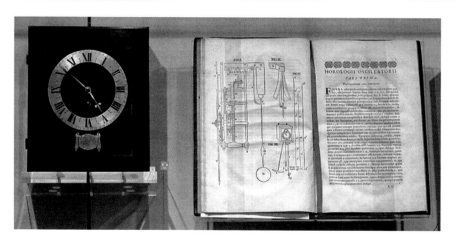

FIG. 7.3 Huygens' pendulum clock and horologii oscillatorii on display at the Museum Boerhaave in Leiden, Netherlands (Image courtesy Rob Koopman)

cans that act effectively as wheels. Mechanically, the beat of met-ronomes cannot be identical; as analog mechanisms, they are sim-ple devices, imperfectly calibrated and regulated. However, once in place on the cans, soon they all lock together in perfect synchro-nization!

The combined motion of the group is increasingly reflected in one stronger motion of rocking back and forth of the platform, as the pulses of those metronomes closest to the final synchroniza-tion begin to dominate, enabled by that 'fluid' coupling of the moving platform. The action of the *singular* back-and-forth motion of the platform lengthens and shortens the weaker individual strokes of respective metronomes, gradually bringing all of them into compliance, when a strong, increasingly combined single cycle regulates the entire group into perfect synchronization, and a strengthened force of motion. Effectively, if not precisely, there has been an amplification of five times the 'forcing' effect of a single metronome acting alone. Once taken off the 'platform,' the beats of the metronomes rapidly move apart, quickly ticking out of synchronization again.

When one takes into account the reported increasing irradiance of the Sun since 1970 (according to the ACRIM research and others), it seems not so difficult to imagine, at least, some additional amplification of solar output occurring through combined planetary gravitational interactions occurring in concert along the lines that Ching-Cheh Hung proposed. Is it entirely inconceivable that this might have resulted in temperatures on Earth being higher than just the expected fluctuations in solar radiance alone would suggest? Perhaps this is why certain periods of solar activity (such as has been theorized for the Medieval Climate Optimum, or the Maunder Minimum) have been more or less active than others. If it could be shown that they occurred when there was a convergence or divergence of planetary tidal forces, and most strongly compounded one way or the other when all forces act together, this could open up entirely new areas of research.

Interestingly Burroughs, as early as 2001, seemed to be leaning in this direction, even if not, perhaps, having become a fully-fledged advocate of such a driver of climate. He did, however, seriously entertain the prospect of such influences on climate in his comments about a number of well-known long-term cycles. Striking was Burroughs' discussion of the possible combined tidal influences of the larger planets, especially, again, with reference to the 179-year alignment on one side of the Sun (see again the section on the Sun's barycentric orbit), so he was already looking at much the same argument. In this regard he mentioned such planetary positioning showing a clear correspondence with the Chinese climate record. If we bear in mind that his book was originally written more than a decade ago, the inclusion of such reasoning is remarkable and is surely a great testament to his open mindedness in looking at all possibilities.

The astronomical component remains, thus, one of the most fascinating and compelling propositions we are likely to find. However, to make any case conclusively depends on the range of inputs and perspectives one brings to the analysis. If there is anything to the possible existence or influence of any of these cycles, surely some will try to prove the possibility and determine their influence.

The Search Widens

Another area of research, somewhat more radical, resulted in a stark but thought-provoking declaration that the driver of climate change was, in fact, atmospheric circulation, and that variations in solar activity were responsible [5]. Along the same lines, and even more unusual, according to another study [6], atmospheric circulation changes happen when 'zonal' circulation is transformed into 'meridional' circulation by the direct effects of solar interaction. Their conclusions were interesting, providing more grist for the mill:

- *Zonal* – occurs during odd cycles and leads to warming, appearing in latitudinal zones.
- *Meridional* – occurs in even cycles and leads to cooling, appearing longitudinally, from north to south (see Rossby Waves, Chap. 3) (Fig. 7.4).

Thus Kuznetsova's group proposed that solar activity in the twentieth century had slowly altered the predominantly meridional atmospheric circulation to zonal, and so the climate had grown warmer. It is interesting that these various researchers did not give any weight at all to any other possible causes! However, it was solar activity, nevertheless, that was indicted as a possible driver of climate variability.

Without jumping to a conclusion that much of the world scientific community has it all wrong, however, perhaps some of the various alternate hypotheses could hold some of the 'missing links,' especially to uncovering some of the subtle contributing factors that have been hard to pinpoint and quantify that commonly have appeared to be 'chaotic.' Perhaps they could help to find a position within the larger, more conventional theories that could be supported by all. Naturally, when there have been such new possibilities raised, it would seem incumbent on science to follow through. Should any alternate theories gain wider credibility or acceptance, if even the staunchest AGW supporter were to continue to claim that all "the science is in," it would seem an increasingly untenable position to take. The onus now is clearly to *disprove* these theories if we ever hope to calm the debate.

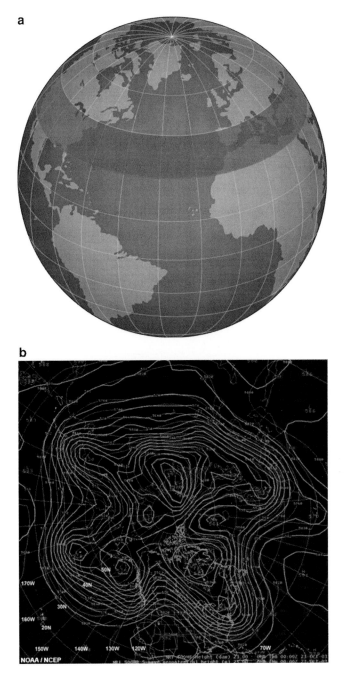

FIG. 7.4 (a) Zonal circulation (Image courtesy U. S. government). (b) Meridional circulation (Image courtesy NOAA/NCEP)

References

1. Duffy PB, Santer BD, Wigley TML (2009) Solar variability does not explain late-20th century warming. Phys Today 62(1):48–49
2. Ching-Cheh Hung (2007) Apparent relations between solar activity and solar tides caused by the planets. Report NASA/TM-2007-214817. Glenn Research Center, Cleveland
3. Niroma T (2007) One possible explanation for the cyclicity in the sun. http://personal.inet.fi/tiede/tilmari/sunspots.html
4. Leubner IH (2007) Variability of stellar and solar radii and effect on planetary orbits and temperatures. American Geophysical Union Fall Meeting 2007
5. Georgieva K, Kirov B, Bianch C (2005) Long-term variations in the correlation between solar activity and climate. Soc Astron Ital 76(4):965–968
6. Kuznetsova TV, Tsirulnik LB (2004) Solar activity cycles in interannual global and hemisphere temperatures on the earth's surface. In: 35th COSPAR Scientific Assembly, Paris, France, July 2004

8. Climate Models

Far from fulfilling their promise of providing clarity of cause and effect that all scientists could agree upon, climate models instead have taken a place at the very core of the debate. The concept is not in dispute; rather, it is the type and quality of the input that continues to fuel the disagreement. However, their importance cannot be underestimated, and as such, they are the result of research closely tied to astronomical monitoring and space studies.

> Models show significant and increasing skill in representing many important mean climate features, such as the large-scale distributions of atmospheric temperature, precipitation, radiation and wind, and of oceanic temperatures, currents and sea ice cover. Models can also simulate essential aspects of many of the patterns of climate variability observed across a range of time scales .... Models' ability to represent these and other important climate features increases our confidence that they represent the essential physical processes important for the simulation of future climate change.
>
> – IPCC Report on Climate Change (2007)

> Since I am no longer affiliated with any organization nor receiving any funding, I can speak quite frankly .... As a scientist I remain skeptical .... The main basis of the claim that man's release of greenhouse gases is the cause of the warming is based almost entirely upon climate models. We all know the frailty of models concerning the air-surface system.
>
> – Dr. Joanne Simpson, U.S. Senate Minority Report 2009
> (Updated from 2007 report)

A. Cooke, *Astronomy and the Climate Crisis*, Astronomers' Universe, DOI 10.1007/978-1-4614-4608-8_8, © Springer Science+Business Media New York 2012

The Advent of Climate Models

Climate models have only been on the scene to any significant degree since the age of solar satellite reconnaissance, just a little over 30 years before the time of this writing. Since then many projections have been made and revised, and just as many changes in methodology have been applied to them! Even the staunchest advocates give little credence to those models created before modern supercomputers were applied to climate. Therefore, that timeline of 30 years would probably be considered quite a stretch by most. So although some of the earliest examples of global general circulation models (the most developed climate model form) date back to the 1960s, probably none of these would be of much, or indeed, any value, today.

Because weather forecasting has always been related to climate, unquestionably the study of both has helped in farming, travel, flood preparation, fires, blizzards, or hurricanes, etc. However, these are short-term scenarios. With climate change becoming a global scientific-political-socioeconomic issue, projecting what the future would hold assumed new and rapidly increasing significance. A growing international alarm developed about the potential for runaway greenhouse conditions encroaching ever-faster, and computer projections (models) became the foundation for climate research. Many climate scientists as well as governments had by now become increasingly convinced that the potential for disaster made it imperative to neutralize additional CO_2 forcing immediately. The downside was that politics had now entered the debate.

With the earliest computers – which filled large rooms and provided less computing power than a modern PC – a new world of potential could be seen to be within reach in the not too far distant future. However, the rudimentary power of these early machines was soon found to be insufficient to handle the multitude of intricate factors involved in replicating weather and climate forecasts with any degree of accuracy. Making realistic forecasts of weather just a few days ahead was enough of a challenge (and frequently a failure), let alone projections for the long term.

In Chap. 3 we discussed time series, Fourier-transforms and power spectra, useful tools to inspect and analyze various factors, but with the element of time removed. All of these, when applied to climate, have certain common ties to climate modeling. In their relationships to climate models, however, these analytical tools part company early and fulfill entirely different functions, but remain standard studies in isolation. They cannot project manifestations of multiple factors on climate or weather on the world globe (or part thereof), nor can they anticipate future responses to other projected inputs. More specifically, they cannot reveal how changing factors will interact with the whole. Again, they are not forecasting tools, but can greatly aid in the understanding of which key inputs have affected observed outcomes.

The tangled code that determines the makeup of Earth's subtle and extremely complex cocoon separates us from immediate extinction, were we to irreparably damage or lose it. For life to continue to exist, conditions inside that cocoon must remain conducive to it, and thus significant temperature change constitutes damage. Computing what had been added to it during the twentieth century and into the early twenty-first century was not difficult, once the Sun's effects had been deduced and separated from pre-industrial age greenhouse warming. With a noticeable increase in temperatures worldwide having taken place over a sustained period of decades, scientists began to theorize how and where the total energy equation might be out of balance, and what might be responsible.

However small this amount of warming may be, it is the key to everything. The problem has been in agreeing on precisely what is happening, what is the cause, and how reliable predictions might be made about the future. Thus, the advent of climate modeling, an attempt to put all known factors together 'in the hopper' and measure their present and future interactions. The total accumulation of anthropogenic global greenhouse gases (GHGs) over the past century, and specifically carbon dioxide (CO_2) above all others, has become central in building these models. Because, rightly or wrongly, climate model scenarios reflect changes to this one factor, virtually to the exclusion of other possible major considerations, they have been challenged for projecting future climate. We will return to this in the next chapter.

The term *anthropogenic* refers to the result of human activity upon the environment. First coined by nineteenth-century Russian geologist A. P. Pavlov, it has, however, become more generally associated with the early twentieth-century British ecologist Arthur Tansley, with his references to 'climax plant communities,' areas that had reached a balance over time, a stable and steady state of existence. These now were undergoing change as a consequence of human influences.

As a result, passionate disagreements about how climate will change in the short and long term have emerged. Although projections have been made from a number of perspectives, all of them are umbilically tied to CO_2. Not everyone has embraced the findings that include proposed remedies, such as keeping future increases of CO_2 to a minimum by a number of methods, or even decreasing levels by fundamentally changing our lifestyles. However, all of this has spawned a counter-discussion at least as loud from those who contend that human activities have not caused the observed changes to the climate, or may have only partly contributed to them in ways that ultimately will not prove detrimental.

At the outset, however, it is fair to say that climate models were never seen as likely hot-button issues, especially of the kind that would cause contentious debate such as we have seen. This is perhaps more the result of the frequent revisions and inaccuracies, variety of projections and predictions over the years, which have resulted from these models being, in essence, works in progress. Past inaccuracies of those from the early days have only become ever more starkly contrasted against more modern projections. Thus when any statistic from even just a *few* years ago is used as part of a prediction, these findings are immediately called into question. However, none of this has stopped many public figures from stating as certainties the scenarios painted by any particular climate model projection at the time of its making, something usually even their designers have shied away from.

Although the IPCC considers that the various models presently in use around the world have a 'considerable degree of agreement,' with each other, this is not quite the same thing as saying

IPCC Definitions of Degrees of Confidence and Likelihoods

Very high confidence	Virtually certain
At least 9 out of 10 chance	*99% probability*
High confidence	Extremely likely
About 8 out of 10 chance	*95% probability*
Medium confidence	Very likely
About 5 out of 10 chance	*90% probability*
Low confidence	Likely
About 2 out of 10 chance	*66% probability*
Very low confidence	More likely than not
Less than 1 out of 10 chance	*50% probability*
	About as likely as not
	33–66% probability
	Unlikely
	33% probability
	Very unlikely
	10% probability
	Extremely unlikely
	5% probability
	Exceptionally unlikely
	1% probability

they are completely consistent or reliable. Therefore it cannot be stated with certainty that even the most advanced climate models have created accurate scenarios for the future. Since all of them share the same philosophy of presumed cause and effect with similar types of inputs, some general agreement of outcome would be expected. However, because all climate models are computer programs that process input according to mathematical formulae, it does not necessarily translate that the less precisely determined and chaotic nature of the universe will necessarily oblige. For long-term models, thus, we may need to wait many years to see if their projections are borne out. Naturally there are many differences of opinion about how well observations have matched existing projections or actual predictions, but this all depends on the observation and model one cites.

Since actual quantifiable measurements of cause and effect have been harder to establish than the projections painted by the models, opponents of anthropogenic theories have continued to blame the use of models for the perception of crisis. A common criticism is that there has been a predetermined assumption among modelers that human activity is primarily responsible for recent climate change, and that Earth cannot respond other than in a negative way. Thus, they claim that worst-case scenarios are just that, and therefore climate models have been designed and modified to reflect this bias.

Weather and Climate

Needless to say, the development of both weather forecasting and climate modeling has steadily continued. Aside from some obvious characteristics exclusive to one or the other, the primary difference between the two remains time, detail, and scale, in degrees depending on the purpose. Weather forecasts work with local weather systems and thunderstorms, local topographical influences on air currents, local oceanic and atmospheric conditions, and so forth. It is also not possible to include some of these smaller factors in climate models, simply because of the lack of computing power for this degree of fine detail over such enormous timelines.

Weather forecasts can be thus verified for accuracy in the short term, and thus the models utilized are easier to modify and develop. We do not have this luxury with climate models, because they are long-term projections.

The chaotic nature of the environment common to both types of models still makes *precise* forecasting of any kind constantly problematic. However, as computing power and sophistication has grown, weather forecasting has improved substantially, enabling even hourly projections to become reasonably accurate. Thus it was only a matter of time before longer-term climate predictions similarly became more refined. The expanding upper limits of computers to integrate and retain information, and at far faster rates, enabled both forms to develop rapidly.

However, a forecast of just a few days ahead requires literally trillions of calculations. This makes it is easy to see how climate models pose almost infinitely greater problems simply because their timelines are likely to extend over decades. And because of limited time scales, obviously a far higher level of 'resolution' could be employed in weather than climate. However, it is those expressions of 'levels of confidence' in climate models, versus having reliable, readily verifiable predictions that remain a fundamental weakness.

The Makeup of Climate Models

Figure 8.1 represents the findings of the model developed for the IPCC in 2007. It is one of the more recent projections for various parts of the world, even though it does not go further than the year 2000. It is thus more of a reference for cause and effect than it is a prediction for the future, the purpose being to confirm what is behind recent trends while utilizing a wide range of inputs.

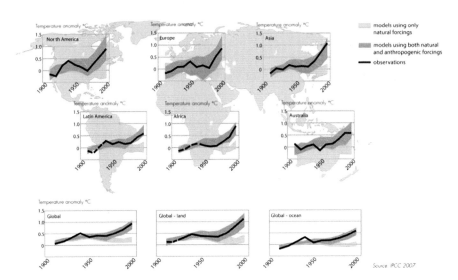

Fɪɢ. **8.1** IPCC climate model projections for different regions (Graphic courtesy of Bounford.com and UNEP/GRID-Arendal. http:maps.grida. no/go/graphic/observed-warming-over-the-20th-century-compared-with-climate-model-calculations)

Thus the future is implied by factors establishing the record as presented. However, charges have repeatedly been made that IPCC models have been regularly altered to reflect observations, and that they have consistently used projections involving a far greater warming effect of CO_2 than has occurred in practice.

The implication again, therefore, is that models have been 'tuned' to fit only the recent increases in CO_2 – and not those earlier in the twentieth century – with the presumption being that only CO_2 could have been responsible. However, it must be said that IPCC graphic model projections such as these may be prone to challenge for at least two other reasons:

- Starting and ending points create an entirely different perspective; IPCC models suffer from too short a timeline at both ends of the projection to allow possible cycles to be represented, especially the 60-year cycle.
- The graphic irregularities have been smoothed over and simplified to a degree that makes almost any projection seem possible and not far out of line.

It could be argued that the visual simplicity of these representations is likely to make the IPCC projections seem more convincing to those who have little background in climate science.

Critics believe further that other possibly stronger factors have received scant attention, through oversights or even deliberate exclusion. Thus they charge that certain significant components of climate have been omitted from the models, or insufficiently weighted in them. These would include the El Nino Southern Oscillation (ENSO), considered by some scientists to be among the principal drivers of climate, and also such possible indirect yet undetermined effects of even slightly elevated solar irradiation. It is no secret that some researchers consider this latter factor alone to be one of the main keys to climate throughout history. However, it is possible to find strongly held convictions on almost anything, as other scientists have taken a contrary stance even on what seemed certain generally agreed fundamentals.

A system of defining subtle varieties of climate model has been developed by the IPCC in order to express concepts more accurately. Reminiscent of some of the fine distinctions between forcings and feedbacks, for our purposes we will try to keep uses of the terms a little more general. Thus, usage of them in this writing may not necessarily coincide with the meanings of the IPCC, those often being hard to separate. The following are excerpts from the IPCC Climate Model hierarchy:

- The basic model is a numerical representation of the climate system based on the physical, chemical and biological properties of its components, their interactions and feedback processes, and accounting for all or some of its known properties.
- A climate prediction or climate forecast is the result of an attempt to produce a most likely description or estimate of the actual evolution of the climate in the future, e.g., at seasonal, inter-annual or long-term time scales. (See also below: climate projection and climate scenario).
- A climate projection of the response of the climate system to emission or concentration scenarios of greenhouse gases and aerosols, or radiative forcing scenarios, often based upon simulations by climate models. Climate projections are distinguished from climate predictions in order to emphasize that climate projections depend upon the emission/concentration/radiative forcing scenario used, which is based on assumptions concerning, e.g., future socio-economic and technological developments that may or may not be realized and are therefore subject to substantial uncertainty.
- A climate scenario is a plausible and often simplified representation of the future climate, based on an internally consistent set of climatological relationships, that has been constructed for explicit use in investigating the potential consequences of anthropogenic climate change, often serving as input to impact models.

Thus, as we have seen before, convolution is no stranger to the IPCC. Meanwhile, depending on the purpose intended, climate models range greatly in inputs, complexity, time frames, regions, type, etc. The late E. N. Lorenz, in the early days of modern climate forecasting, established two principles in the definition of predictability in atmospheric studies. (Lorenz also was one of the first modern scientists to develop theories surrounding the chaotic nature of the environment, and to bring the concept into general usage).

- The most straightforward form of predictability centers on an established base line of factors that are known quantities on which to calculate the response.
- The other principle involves more complex processes concerning the interaction of lesser-known factors together, without a known finite base line on which to calculate the response.

These principles apply no less now than they did when Lorenz first proposed them, and certainly seem very close to the concepts of climate modeling in general. Unknowns – the second principle – remain the single biggest issue and a source of controversy for climate models to overcome, in which adjustments, or 'tunings,' have been performed on them regularly in order to make them fit otherwise non-quantified forcings. Although these may exist in theory, in practice they have been harder to show to everyone's satisfaction. For example, the temperature dynamics of the oceans are far from understood as a component, and continue to pose more questions than answers.

Additionally, the longer the time frame of a climate projection, the less 'finely resolved' information can be included. This means, therefore, that longer-term climate models are primarily concerned with creating estimates of the future climate and overall conditions of large world regions, rather than details specific to smaller areas. This is why, even with today's mighty computers, it is still impossible to project details of even a slightly comparable resolution to those found in weather forecasts. It is also why no one really can say for sure how accurately present models are able to anticipate future climate conditions.

Climate models as a whole can be classified accordingly:

- The simplest are termed Energy Balance Models (EBM's) and are usually considered low-resolution models covering larger world regions with simple inputs.
- More complex atmospheric one-dimensional Radiative-Convective Models (RCM's: radiation balance and heat transport of convection), or two-dimensional Statistical-Dynamical Models (STM's: combination of other energy balances and RCM's together).
- Earth Models of Intermediate Complexity (EMIC's) are generally developed in comparison to EMB's, containing many

more parameters and inputs but still falling below maximum possible parameters.

- General Circulation Models (GCM's) is a general term for models that are built around a mathematically divided spherical grid of Earth and may be utilized to project outcomes from a wide range of inputs and resolutions for any given time frame and/or outcome being considered.

Specific varieties of GCM also exist, depending on the purpose intended and types of included inputs:

- Those that model ocean environments are termed Ocean General Circulation Models (OGCM's). These are used to predict future temperature patterns, salinity, currents, etc., through the entire depth of the oceans.
- When modeling atmospheric factors, the models are known as Atmospheric General Circulation Models (AGCM's). In these, the influences of wind, precipitation, atmospheric temperature, humidity, etc., are included.
- Coupled General Circulation Models (CGCM's) utilize all factors linked together and also may be termed Atmosphere Ocean General Circulation Models (AOGCM's). These are the most elaborate and inclusive types of models.

To build a model, one needs to divide the globe (or portion thereof) into a format of workable units within a three-dimensional grid, each being usually a maximum of a few hundred square kilometers in dimension. As time-dependent units these are collected over a finite duration, in specific volumes according to the division of the three-dimensional grid. Figure 8.2 shows the general concept of such a model, but limited to the atmosphere (an AGCM).

Having thus divided the model sample into symmetrical cells by volume, the behavior of each and the responses over time are computed as mathematical equations, each being compared against adjacent cells. In this way, an overall global model is assembled. Unknowns include whether/how Earth will respond to increased forcings of anthropogenic origin, whether it will effectively neutralize them, or something in between – maybe even none of these. Climate modelers have made certain determinations according to theory and past observations, the gray areas that are central to the ongoing debate.

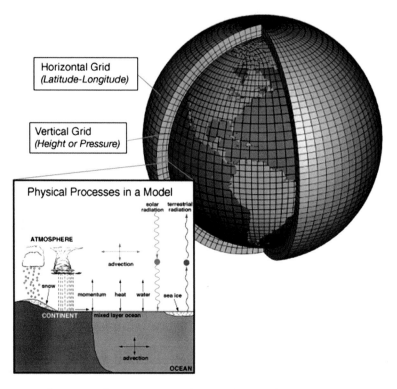

Fig. 8.2 Schematic for an atmospheric global climate model (Graphic courtesy of NOAA)

However, we should examine further the most sophisticated model type in a little more detail (AOGCM or CGCM), the type used in most of today's total global climate projections – especially since the IPCC considers them to be essential components in their periodic assessments. In an AOGCM, full measures of all known (or at least those that are acknowledged) interactive components are accommodated, including but not limited to:

- Known forcings and feedbacks, and compounded feedback loops.
- Biological and chemical factors and aerosols.
- Atmospheric currents, convection and pressure.
- Water vapor and cloud cover (the latter a subject of dispute as to whether it is a positive or negative forcing; all current models used by the IPCC project them as positive).
- Winds, wind speeds, and directions.
- CO_2 storage and recirculation, including storage and release from land and ocean heat sinks.

- ENSO and NAO oscillations and other air and ocean currents.
- Solar variations, and radiation input and output at all latitudes.
- Deforestation, changing albedo.
- Variations in snow and ice cover, water density and composition, according to speed and degree of ice melt, further changing albedo.
- Responses of greenhouse gases to Earth's emissive radiative wavelengths.
- Computed energy loss to space, after all factors are taken into consideration.
- The interaction of all of these factors together.

After this, each long-term model has to be adjusted according the projected seasonal changes in weather patterns, as well as differentiating between all global regions. It may need to cover decades into the future. Because so much of Earth's weather and climate is chaotic in nature, this makes precise predictions and projections further in advance increasingly problematic. We have already seen how certain longer-term influences on weather can ape true climate change (see again Chap. 3).

Models of all types and complexity may even be assembled for specific regions in isolation, having entirely different contributing forcing and feedback processes than the greater surrounding areas. High-resolution three-dimensional modeling applied to limited regions and time scales have been able to simulate ocean eddies and show the flexibility of this technology. Because ocean eddies have a great influence on the mixing of waters – and ultimately on deepwater circulation – they influence climate more than might seem obvious at first glance. In this instance, Fig. 8.3, an OGCM, the specific modeling of eddies in the northern Atlantic Ocean, shows that precise information may be deduced.

Figure 8.4 illustrates modeling on a global scale, at a 10–12 km resolution. Although limited to the study of cloud movements and circulation only, it covers the entire globe. We can consider it an AGCM, although this is again a highly short-term projection. In this instance NASA created a moving graphic for a 1-day period, from which this is excerpted.

Meanwhile, Fig. 8.5 shows something we may be far more familiar with: a coupled climate model for the entire global surface to project future temperatures for the remainder of the twenty-first century. This is far more typical of the climate models we are

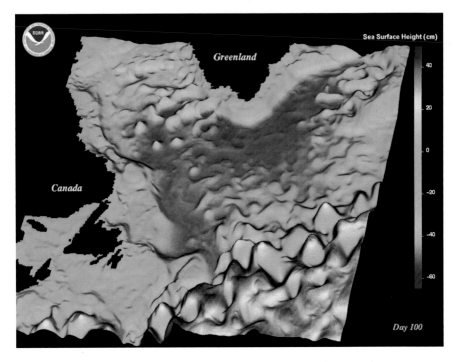

Fig. 8.3 Eddies off Greenland and Canada in the northern Atlantic Ocean (Image courtesy of NOAA/GFDL)

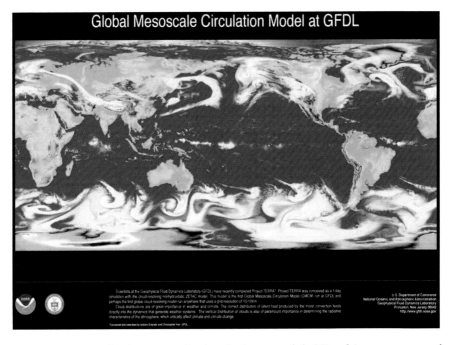

Fig. 8.4 Atmospheric general circulation model (Graphic courtesy of NOAA/GFDL)

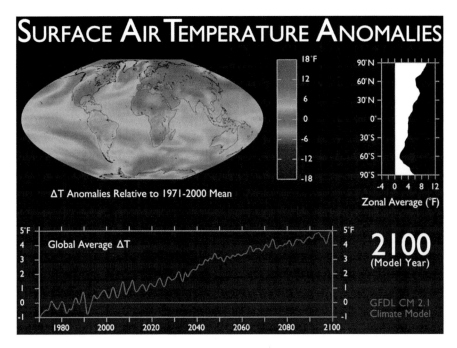

SURFACE AIR TEMPERATURE ANOMALIES

ΔT Anomalies Relative to 1971-2000 Mean

Zonal Average (°F)

Global Average ΔT

2100
(Model Year)

GFDL CM 2.1
Climate Model

Fɪɢ. **8.5** 2005 (CGCM) 1971–2100 (Keith Dixon) (Graphic courtesy of NOAA/GDFL)

likely to see in the mainstream. It is not comforting, although we must remember it is only a projection based on acknowledged and assumed factors. However, it is precisely the drastic conclusions of models such as this that have been the target of so much criticism by those who maintain that the wrong forcings have been applied or omitted, with entirely unrealistic projections being reached.

Water Vapor and Clouds

The subject of clouds and water vapor are also at the core of the various controversies in climate models, as well as trying to determine whether they function first as feedback or forcing. Cause and effect of both water vapor and clouds are still far from understood or agreed, also in respect to which came first – the warming or the clouds – and whether heat is lost because of them or gained. Water vapor is to some degree self-regulating through precipitation back into liquid water. What is not so clear is how warming affects its overall impact on climate; as an intensely complex and transitional component it is insufficiently understood. Thus, its effect must be largely assessed depending on one's faith in computer projections and estimates.

As one of the central tenets of climatology, the ongoing study of these hydrological factors has a number of well-known followers, such as climatologist Roy Spencer. His position and theories about climate models can be accessed on his own website in addition to links on his various peer-reviewed research papers [1]. However, Spencer is no stranger to controversy, and his positions are obviously at odds with other practitioners of mainstream science. He remains a prominent and respected climatologist regardless, believing instead that Earth has many built-in systems of self-regulation. Apparently this is another manifestation of the 'Gaia Hypothesis' (see Chap. 1), albeit in an unsung capacity.

Climate Model Errors and Questionable Temperature Readings

Because numerous researchers remain steadfast that present conditions could not have originated with present allowed inputs only, their charge is by now familiar. In short, key factors have been omitted, while others, notably the influence of carbon dioxide, a measurably increasing *known* quantity, has been exaggerated in order to make the models work.

The Accuracy of Climate Models

According to the website Global Warming – Man or Myth, it is easier to predict climate than it is weather. If this were so, the argument would already be over, and there would be nothing to write or talk about. Meanwhile, few devout supporters even of the IPCC position would be likely to make such a wild claim, in light of the constant revisions found necessary in the long and arduous evolution of climate models.

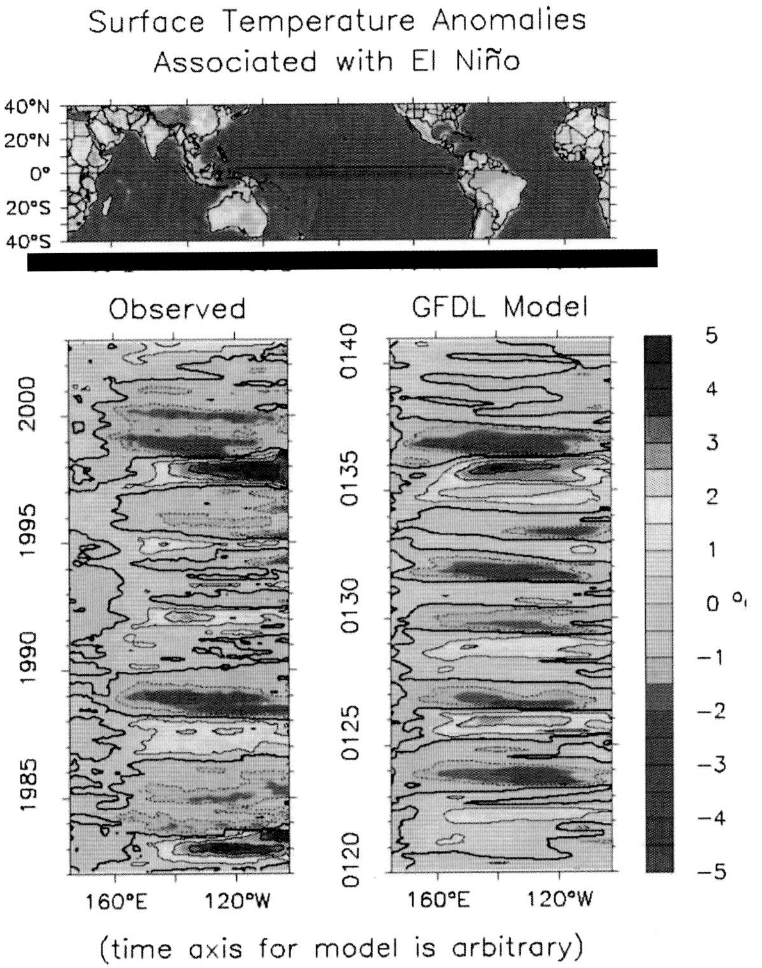

FIG. 8.6 ENSO coupled climate model (Hovmoeller) and parallel observations (Image courtesy of NOAA)

Although it seems no scientist or research center has claimed to have 100% faith in these models, their proponents believe that overall they are reasonably accurate – or to paraphrase the IPCC, that there is a 'high degree of confidence' in them, such a lack of certainty being the constant bone of contention. Skeptics have questioned the prospect of making major changes in the way the people live, against uncertain, or even unlikely, projections of the future. Certainly, it is not hard to find simple errors in past model projections. Figure 8.6 represents the NOAA/GFDL record since 1980 in modeling for future ENSO/El Nino events. Comparisons

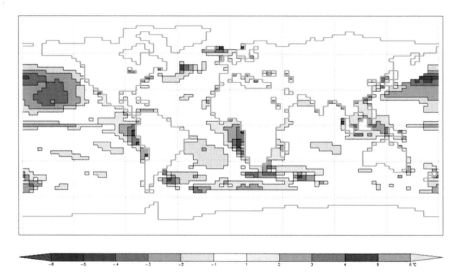

Fɪɢ. **8.7** Errors SST in Hadley Climate Model 3 (Graphic courtesy of William M. Connolley)

of observation against model are instructive, as the resemblances often are none too close.

Although it can be seen in this particular example that the accuracy between 1985 and 2000 appeared to have improved somewhat, the prediction of events was still far from precise. In defense of those trying to develop such predictions, it must be realized that precise definitions of any factor in projection will, by default, always be subject to error, and small errors can indeed create entirely different outcomes. However, a strong and theoretically demonstrable component of the skeptics' argument has been that if just one forcing effect has been overestimated, that this alone will completely negate all the outcomes projected by climate models.

Figure 8.7 illustrates precise levels of error over a 1-year time span for sea surface temperatures (SST's). Projections of the OCGM Hadley Climate Model 3, and actual observed temperatures (1978–1994), were reviewed for correction in the re-analysis project ERA-15. The errors are not inconsiderable, and are displayed in degree according to color. They underline the difficulties faced by climate modelers, as well as the potential for skeptics to call the entire methodology into question. We should remember, however, that the models used in this analysis date from some time back; it is unlikely they would be considered valuable by today's standards, regardless.

Satellites have become increasingly important in supplying more reliable temperature readings to climate modelers; their reliability is increasingly trusted. Indeed, their accuracy today is virtually universally agreed and accepted, so at least there is one area of climate science that is not in dispute. A notable study in *Nature* reaffirmed not only the degree of accuracy of satellite readings versus other methods of temperature measurement (i.e., balloons) and climate models, but declared that lower atmosphere temperatures had actually declined since satellites had been utilized some two decades earlier [2]. However, this is a diametrically opposite stance to NASA's own published records for 1979–2011 (see again Fig. 5.8, Chap. 5).

Although climate models had projected significantly higher temperatures through increases in greenhouse gas, the *Nature* study, at least, did not find that to have been an accurate outcome. However, before all skeptics run a victory lap, it should be pointed out that this article dated from 1997, the year before the 1997/1998 El Nino spike, and well before the NASA graph in Chap. 5. At very least, however, even if it did not address the current situation, it seemed to confirm that the models up to that time were wrong. Regardless, the same factors of anthropogenic gas increases are still being included in present climate models. This apparent contradiction has not yet been laid to rest to everyone's satisfaction.

Figure 8.8 illustrates the history of all types of measurements combined. It also shows how one can draw different conclusions from the same information, and again seems to fly in the face of the '97 *Nature* article that claimed lower atmosphere temperatures actually had fallen. On this graphic, if we look at the three averaged trend lines it seems clear that all of the trends agree in essence, and that overall surface and lower atmosphere temperatures had *increased* over the period. However, if we reset our starting point after the '97/98 El Nino spike, at least (the time of the *Nature* study) we can draw another determination: there seems to have been no overall warming at all since 1998. It is not surprising, therefore, that this and the reliability of temperature readings (especially surface-based ones) have become central to present arguments on climate change. In such examples, it can depend on where we set our starting and ending points and draw our intersecting lines.

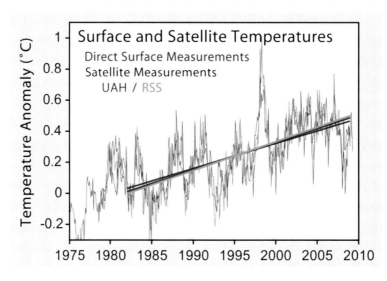

FIG. 8.8 Surface and satellite temperatures combined (Graphic courtesy Global Warming Art Project)

The Urban Heat Island

Still other reservations about the accuracy of climate models stem from the twentieth-century phenomenon called the *urban heat island effect*. It has been observed in areas of significant urbanization that the climate is affected to some degree, with resulting hotter temperatures. The causes of these higher temperatures include reduced areas of vegetation, increased density of roads and buildings, reduced permeability to water (with consequently drier and warmer surface conditions in sunlight), increased storage of heat in buildings and concrete pavement, as well as significant reductions in the amounts of moist areas of wooded shade. The consequences are two distinctly different 'heat island' effects:

- Surface temperatures tend to be somewhat higher during daylight hours than elsewhere, although they continue to radiate heat at night at a reduced rate.
- At night, the air immediately above the urban landscape becomes increasingly warmed from the outward radiation stored during the day in structures below. Thus nighttime temperatures tend to be relatively much hotter than in rural

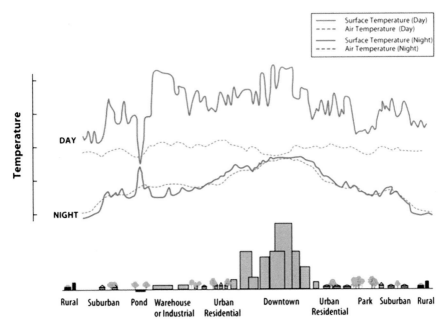

Fig. 8.9 Effects of the urban heat island effect (Graphic courtesy of NOAA)

areas where the ground has far less stored heat near the immediate surface.

The EPA estimates that the urban heat island effect can be as large as 3°C during the day in a city of at least one million people, and a surprising 12°C at night. The great divergence is, of course, due to rural areas cooling down much more than do urban areas, full of heat-retaining infrastructure, during the night (Fig. 8.9).

Another effect within the urban heat island phenomenon can be clearly seen in Fig. 8.10. Interestingly, tall buildings often retain heat to far lesser extent than the surrounding urban sprawl, presumably because heat is more readily radiated from the more isolated taller structures than those near ground level, and the fact that most of them are glass-covered and highly reflective. When this image of Atlanta was made in 1997, the actual temperature recorded in outlying areas was approximately 80° F, but within the city, temperatures as high as 118°F were recorded!

The reason this phenomenon is so important is that it is connected to the way global temperatures are collected and inputted

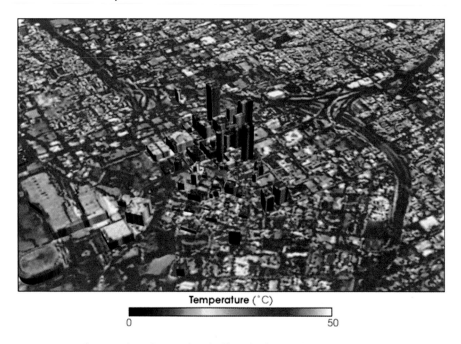

Temperature (°C)

0   50

FIG. 8.10 Atlanta urban heat island effect (infrared) (Image courtesy of NASA)

into climate models. If the numbers being fed into the models do not reflect accurately what is really happening, the models will reflect and possibly compound such errors in computations many times larger. Critics have continually raised questions about the reliability of temperature measurements used in models, whether due to the heat island effect or the various other methods utilized for temperature collection. Among the criticisms:

- Many monitors/sensors are located within areas of urban heat islands.
- Early models did not accurately account for the urban heat island effect.
- There is a greater density and longer history of land-based temperature sensors located in the United States than elsewhere.
- Ocean sensors are carried onboard buoys, with varying reliability. Plus, early sensors were on board ships, and overestimated SST by 0.6°C because of the heat from the ships' engines – another related effect.
- Satellite SST readings may be unreliable because they usually monitor only the tops of the oceans' surfaces and not immediately below, where truer readings lie.

- Satellite measurements cannot penetrate cloud, leading to underestimations of temperatures, although microwave technologies have now helped to remedy this defect.

The last two reservations seem not to instill confidence that the satellite record can always be trusted. Therefore, despite claims that everything has been taken into account, and thus, despite the great strides in monitoring technology that has been made over the years, it is clear that the complete accuracy climate modelers seek has been demonstrated to few peoples' satisfaction. Even proponents and designers of the models have found plenty of fault with them. Kevin Trenberth, in a 2011 article in *Climate Research*, highlighted some significant shortcomings in climate models, ranging from:

- Poor representation of transient tropical weather systems, including hurricanes and tropical storms.
- Overstatement of frequency of precipitation and understatement of intensity.
- The possibility that all modeling has built a too-rapid convection effect, leading to the overstatement of precipitation.
- Inaccurate simulations of the hydrological cycle in general.
- Too little atmospheric residence time projected for water vapor [3].

Such shortcomings are far more significant than everyone might realize, especially since water vapor plays such an important part in the entire climate, not the least of which is its huge greenhouse contribution. Critics have argued if this one key aspect cannot be accurately represented, climate models are not likely to be trustworthy. We should remember that just small inaccuracies in these models have been implicated in having far reaching effects on the accuracy of overall projections.

There is no shortage of questions about the methodology used in climate models and the quality of their projections. Researchers R. Fildes and N. Kourentzes cast serious doubts about the accuracy of even a 10-year projection [4]. However, in their study they proposed a way to reduce the controversies by blending time series and climate models together. They maintained that their method would produce far greater reliability, and therefore ultimate acceptance of these models. However, in a bold position, perhaps risking outright rejection, the authors stated that they considered efforts to regulate carbon emissions above all else had been "misguided."

Another finding from Oregon State University was that climate models had probably based future carbon dioxide release from soils at latitudes too high and too low, depending on variable temperatures in each region [5]. Terrestrially stored CO_2, though a far smaller 'sink' than that of the oceans, was actually responsible for *more* net annual exchange. Because the total annual release of all fossil fuels is only 5.4% of all CO_2 exchanged by the atmosphere, the potential was considerable for even small errors (in estimates of carbon release from the land sink) to add to or offset all or most of the fossil fuel projection. It was not hard to deduce that probably all net total added CO_2 represented not more than 1.5% of the total amount exchanged each year. Compare this again with the author's deduction in Chap. 2, which put this amount in similar territory, at 0.93%.

If the range of challenges and suggested remedies seems endless overall, however, the creators of climate models can justly claim considerable success in improving their projections in recent years. For those who maintain that new considerations have not been factored in, and that their proponents have been shut out of discussions taking place among those who design them, unfortunately, these models also have served to ensure that the divide between both sides of the debate stays energized.

References

1. Spencer R (2009) How do climate models work? http://www.drroys-pencer.com/2009/07/how-do-climate-models-work/
2. Christy JR, Spencer RW, Braswell WD (1997) How accurate are satellite 'thermometers'? Nature 389:342–343
3. Trenberth K (2011) Changes in precipitation with climate change. Clim Res 47:123–138
4. Fildes R, Kourentzes N (2011) Validation and forecasting accuracy in models of climate change. Int J Forecast 27:968–1005
5. Sierra CA, Harmon ME, Thomann E, Perakis SS, Loescher HW (2011) Amplification and dampening of soil respiration by changes in temperature variability. Faculty Research Publications, Oregon State University, April 2011 http://www.ir.library.oregonstate.edu/xmlui/handle/1957/21537?show=full

9. Interpretations of the Data

Despite the earnest and best efforts of scientists, the sheer complexity of the subject makes precise agreement between studies, types of research, quality and interpretations of the data subject to any number of differences and disagreements between colleagues. The passion that climate science has generated has sometimes tended to make these become personal, especially when reputations are on the line and hard sought research has taken years of dedicated work. As such, climate science has not been without its star players, who have found themselves either embraced or becoming targets of the opposite side.

James Hansen, head of NASA's Goddard Institute for Space Studies (GISS), and perhaps the most famous climatologist in the world today, has headed some of the most frequently quoted research findings that leave little doubt that CO_2 is the primary culprit of recent climate change [1]. Hansen certainly is supportive of the 2007 IPCC position and has, of course, been one of the leading advocates for the theory of anthropogenic causes of climate change for decades. As such, he is no stranger to criticism, some of it far from objective. His views represent the many aspects of climate studies that can be supported by space research through GISS, his hallmark being never to deprecate the efforts of other researchers of any stripe. Perhaps surprisingly, various papers that Hansen headed show statistics that seem to contradict themselves, for example, discrepancies in temperature readings over the years:

From 1981 [2]:

- The 1981 measurements put global air surface temperatures at their highest in 1940 – at just over 0.4°C higher than in 1900 and at almost 0.1°C less than in 1980. Then the temperature gradient follows a jagged decline until about 1970, increasing after that time.

A. Cooke, *Astronomy and the Climate Crisis*,
Astronomers' Universe, DOI 10.1007/978-1-4614-4608-8_9,
© Springer Science+Business Media New York 2012

"A remarkable conclusion is that the global temperature is almost as high today as it was in 1940."

From 1987 [3]:

- Without apparent explanation, another graph of measurements from 1987 shows that suddenly the high temperature in 1940 now registers almost 0.1°C less than shown in 1981, while that of 1980 registered higher by the same amount. The accompanying quote is strikingly at odds with that of just 6 years earlier.

"The global air surface temperature in 1981 reached a warmer level than obtained in any previous time in the period of instrumental record."

From 2009 [4]:

- Recent new graphs showed the temperature in 1900 had been lowered by 0.1°C. Although that for 1980 had also been reduced by the same amount, the steep slope up to the high temperature in 2007 (0.55°C) was even more drastic.

> The analysis method was documented in Hansen and Lebedeff (1987), showing that the correlation of temperature change was reasonably strong for stations separated by up to 1,200 km, especially at middle and high latitudes. They obtained quantitative estimates of the error in annual and 5-year mean temperature change by sampling at station locations a spatially complete data set of a long run of a global climate model, which was shown to have realistic spatial and temporal variability.
>
> This derived error bar only addressed the error due to incomplete spatial coverage of measurements. As there are other potential sources of error, such as urban warming near meteorological stations, etc., many other methods have been used to verify the approximate magnitude of inferred global warming. These methods include inference of surface temperature change from vertical temperature profiles in the ground (bore holes) at many sites around the world, rate of glacier retreat at many locations, and studies by several groups of the effect of urban and other local human influences on the global temperature record. All of these yield consistent estimates of the

approximate magnitude of global warming, which now stands
at about twice the magnitude that we reported in 1981.

– NASA

Before we rush to judgment that someone has been playing
fast and loose with the facts, however, we should bear in mind
that these results only reflect the thinking at the time of the study.
They do not represent the inputs of any one individual or institution, although certainly those of the institution (GISS) that Hansen heads are the major component. Regardless, it is easy to see
why some have questioned the methodology utilized.

But this will probably not satisfy everyone. Four issues come
to mind at first glance:

- The last reference to using a 1987 climate model is highly questionable. Even now, newly formulated models are far from universally accepted as being reliable tools.
- The reference to the 'urban heat island' effect in paragraph 2
has no relevancy for ocean temperatures, which, at 70% of
Earth's surface constitute a large proportion of the global readings. Therefore, if any adjustment were necessary, we would
assume that the adjusted global readings for 1980 from the earliest report (1981) would have registered too *high*, instead of
too low.
- Similarly, because the actual occurrence of the urban heat
island effect was even less in 1940, one would think that those
adjusted readings would have been too low, not too high!
- On the other hand, measurements of ocean temperatures also
present their own set of reliability issues, especially since these
readings are often taken from moving vessels and are thus subject to other adjustments [5]. (See also Chap. 8.)

Regretfully, NASA's explanation only helps in our quest
for clarity and consistency *if* we have already accepted their position – a good example of the problems we have raised. Justly or
unjustly, the various scenarios Hansen painted reveal a history of
constant tinkering over the years. Two related graphs (for U.S.
temperature) again readily illustrate the essence of the problem
(Fig. 9.1):

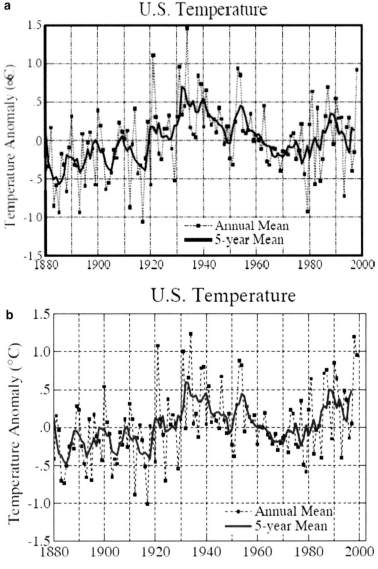

FIG. 9.1 (a) An actual temperature graph by Hansen's team (GISS) for the United States, dating from 1999 [6]. It shows a clear downward trend in temperature from a peak in the 1930s. (b) Compare (a) carefully with a parallel plot by Hansen's team from 2001 [7]. Though superficially similar, we can readily see that subtle changes make it appear that temperatures have been on an upward trend since the 1930s – totally opposite to the conclusion from just 2 years earlier!

Although these adjustments were justified according to stated newer enhancements in measurement techniques, etc., regardless of their validity, reassurances that only favor one's case do little to calm controversies from staunchly distrusting critics. More problematically, a critical part of the argument has been brought up, in that apparently CO_2 had no effect on climate prior to 1980, but since then it has. As you will recall, critics have pounced on this with the accusation that CO_2 was indicted only to make climate models work, because there was no other way to explain the rise in temperature. Indeed, NASA has made this very justification. Make of this what you will.

James Hansen maintains that advancing techniques have allowed more accurate determinations of statistics from past records, which surely is an accurate statement of his convictions. Hansen's perspective can better be appreciated, unfiltered, from excerpts of a recent press release from the Goddard Institute (GISS):

NASA Research Finds 2010 Tied for Warmest Year on Record
January 12, 2011

Global surface temperatures in 2010 tied 2005 as the warmest on record, according to an analysis released Wednesday by researchers at NASA's Goddard Institute for Space Studies (GISS) in New York.

The two years differed by less than 0.018 degrees Fahrenheit. The difference is smaller than the uncertainty in comparing the temperatures of recent years, putting them into a statistical tie. In the new analysis, the next warmest years are 1998, 2002, 2003, 2006, 2007 and 2009, which are statistically tied for third warmest year. The GISS records begin in 1880.

The analysis found 2010 approximately 1.13°F warmer than the average global surface temperature from 1951 to 1980. To measure climate change, scientists look at long-term trends. The temperature trend, including data from 2010, shows the climate has warmed by approximately 0.36° F per decade since the late 1970s.

The record temperature in 2010 is particularly noteworthy, because the last half of the year was marked by a transition to strong La Niña conditions, which bring cool sea surface temperatures to the eastern tropical Pacific Ocean. A chilly spell also struck this winter across northern Europe. The event may have been influenced by the decline of Arctic sea ice and could be linked to warming temperatures at more northern latitudes.

The loss of sea ice may also be driving Arctic air into the middle latitudes. Winter weather patterns are notoriously chaotic, and the GISS analysis finds seven of the last ten European winters warmer than the average from 1951 to 1980. The unusual cold in the past two winters has caused scientists to begin to speculate about a potential connection to sea ice changes.

Within the scientific community, it is possible to find many examples of entirely different conclusions theoretically based upon the same evidence. However, it is not necessarily true that all, let alone *any* of the recent, seemingly esoteric astronomical concepts have been taken into consideration, especially since most researchers cannot be intimately involved in every science. Consequently, many positions might have been made in a partial vacuum. Regardless, the field of climate study has already long established positions and methodologies; many of the more compelling astronomical findings and theories are relatively new on the scene.

Empirical Evidence for the Effects of Carbon Dioxide

The crux of the entire climate change controversy, the largest single focus of disagreement of all, is centered on carbon dioxide, CO_2. The difficulties of pinning down its effects have ensured that it remains the largest single data anomaly.

In the twentieth century, climate scientists have universally accepted two distinct periods of warming: that which occurred prior to 1970, and what has occurred since. There has been a distinct separation of the attribution of greenhouse warming in the first period and the second. Virtually every climate model has contained the inference that carbon dioxide has been responsible for the second period of warming, because concentrations have been steadily rising with no other possible new contributing factors being incorporated as possible forcing agents.

Thus CO_2 remains central to anthropogenic theories, and central to the formulae to make climate models fit the observed temperatures. However, the problem from the start has been in quantifying the effects of the gas, since many projections have been built on its anticipated effect. The problem is that direct evidence for it is largely missing – that the recent warming and retreat of Earth's ice sheets, etc., has been *indirectly* attributed to CO_2 – and thus all we have is circumstantial evidence at best, and expectations based on physics. However, we have already seen that even the physics regarding the warming effects of atmospheric CO_2 has many interpretations. Perhaps, though, a truer argument would be whether *all* warming is due to anthropogenic causes. After all, it seems that many, if not most, researchers looking for other causes of recent warming have not claimed that their findings *exclude* anthropogenic contributions as part of the mix. In theory at least, increased greenhouse gases should lead to some warming.

Note that most serious critics refer to exaggerations of the effects of AGW, while not necessarily denying them, although that is something that all of them have been incorrectly accused of doing. Although the degree of supposed exaggeration has been variously estimated, and some have certainly considered it statistically insignificant, maybe an approach that holds it *partly* responsible could prove to be one way out of this mess. For any chance of this to happen would mean that everyone concerned would have to be open to any and all possibilities, and all recent studies. Should either side hang onto their position that they understand the entire picture, it would not break the logjam. However, such a scenario would have to be ultimately verifiable, of course, since we should not be looking for a popular vote and a happy medium of opinions to settle science matters, or to impose something expedient. Canute again.

The Linking of Warming to CO_2

The Environmental Defense Fund has declared that CO_2 is definitely behind global warming, and that nations have the responsibility to control future emissions.

The website *Skepticalscience* has stated that direct measurements indicate that increasing levels of CO_2 are trapping more heat, that human-induced CO_2 is responsible for global warming, and known with high certainty and confirmed by observation – that CO_2 is the main driver of climate change and that human CO_2 emissions are what is behind global warming. Similarly, the NRDC, the National Resources Defense Council, claims that unless "global warming emissions" are reduced, average U.S. temperatures could be 3°–9° higher by the end of the twenty-first century. The NRDC did not even provide the reference temperature scale (°C, °F, or K) while quoting unnamed "scientists" to back up their claim.

If we search for actual empirical evidence of the link between recent warming and carbon dioxide, we will find little that allows us such take such bold stances as these. Even those scientists most passionately on-board with AGW theories are careful not to fall into the trap of stating such unequivocally rigid dogma, preferring instead to frame their conclusions in terms more akin to 'likelihoods.'

More specifically, therefore, what scientific evidence exists of that direct link between increased levels of CO_2 and rising temperatures? Since we find so many references from those who do not accept anthropogenic warming theories that there is "no empirical evidence linking carbon dioxide to recent warming," is that, in fact, true? The fact that the debate continues so vigorously to this day would seem to imply strongly that such measurements have not yet been demonstrated in a convincing manner. Otherwise, we would expect doubters to have been silenced.

However, a somewhat definitive-sounding article by John Cook on his Skeptical Science website offers a synopsis of what it can [8]. Satellite reconnaissance and surface measurements have revealed less infrared energy escaping to space at wavelengths associated with CO_2 absorption. That may well be indisputable, although Cook apparently is unable to state any point of reference to the actual degree of warming that has resulted from that. Beyond this, unfortunately, the summary case that Cook presents is not at all conclusive, let alone compelling. Pointing to the fact that the oceans and land have warmed is further supposed to prove that CO_2 was responsible! However, no one has disputed that temperatures have been rising, so we need to examine some of the more salient references provided in the article, because they represent the best efforts of those looking for such a definitive link.

The first two studies cited involved a common author, taking the same approach of hedging bets on specifics [9]. Since neither is particularly recent, we can assume that they were not able to end the dispute. Though primarily concerned with increases in greenhouse gas, the first happened to make mention of the fact that the hydrological cycle – a key factor, if ever there were one (water vapor, rain, ocean currents, etc.) – was not yet well understood. Both found changes to outgoing radiative functions of greenhouse gases, 'consistent with concerns over radiative forcing,' but no measured consequences of temperature increase.

However, careful analysis revealed that the strongest increase in retained heat occurred in absorption bands of a greenhouse gases *other* than CO_2. Specifically, it was methane (CH_4) that scored highest on the totem pole. Methane, incidentally, represents less than 1/22 of the total volume of atmospheric carbon dioxide, and about a quarter the radiative forcing value, molecule for molecule.

A contrasting view concluded that the effects of water vapor were "significantly underestimated," while those of CO_2 were "significantly overestimated" [10]. These are not inconsequential remarks, considering how much weight has been given anthropogenic greenhouse contributions. The author also reminded us that increases in CO_2 levels, historically, were the result of warming rather than the cause, a charge we have heard before from others. He continued to remind us of the uneven rates of temperature increase over the twentieth century versus the near monolithic increases in CO_2 levels. Perhaps most interesting was the projection that CFC and CO_2 heat saturation had probably occurred by the end of the twentieth century, and that a long-term cooling will follow for up to seven decades.

Water Vapor and CO_2

Among those arbitrarily claiming as a fact that CO_2 warming has led to increases in atmospheric water vapor, we will find both the Environmental Defense Fund and the website *Skepticalscience.*

Although such a scenario exists in most theories, CO_2 warming has yet to be demonstrated. In the absence of empirical evidence of actual CO_2 warming (existing as yet only in computer models), it is difficult to determine if there has been any anthropogenic influence resulting in observed increases in atmospheric water vapor.

Other researchers have tried to be specific; one estimate of warming was a global net forcing increase of 2.2 W/m^2 per decade from 1973 to 2008 [11]. However, the trend was attributed to increases in water vapor and CO_2, although it is unclear how that conclusion was reached, or what was used to support it. However, different research put the total energy flux imbalance from all increases in anthropogenic emissions since 1850 at 3.52 W/m^2. This is clearly quite a different conclusion [12]. The abstract of the paper contains a statement that the findings should end the discussion that "no empirical evidence exists to connect increases of greenhouse gases and global warming." Although starting out with a good premise, it was far from successful in substantiating those bold claims that accompanied it.

However, as an important effort to quantify with actual physical evidence of otherwise unsupported claims made by others it is perhaps the best we have. Its greatest weakness may have been in prioritizing the analysis of increases to almost every minor other trace greenhouse gas, most notably CFC's, ahead of CO_2. Thus, the authors did *not* provide a strong case for the link that supporters of AGW are trying to make, namely that CO_2 is the primary driver of climate change. However, the carefully chosen words of the summation seem to disguise what may be an avoidance of the key element of AGW theories – namely, CO_2! Indeed, in the introduction, CO_2 is not even mentioned.

Furthermore, the specifics that study provided, although certainly a commitment to actual numbers, showed:

- An over-reliance of comparisons against simulated fluxes, estimates and formulae.
- Compared to these simulated readings, most readings are not so striking that they could be argued to represent more than just statistical anomalies.
- An insufficient breadth of sampling locations, those utilized having been taken at a single mid-latitude location only.

In fact, in one key table ('measured summer downward surface fluxes'), the actual measured downward radiative flux for CO_2 shows *no change* between 'past' (read, *simulated*) readings and the measurement made in 1999. Although this table attempts to pit readings in 1999 against those of the pre-industrial period, because these records were simulated it surely cannot be claimed the argument has been settled – that is, without the expectation of any challenge.

The many other listed and analyzed compounds are indeed greenhouse gases, but they are considered even by the IPCC to represent collectively at most only 23% of the total warming – excluding water vapor (see Chap. 2, Fig. 2.5). Thus, the purported primary driver of AGW (CO_2) was marginalized, rather than the other way around. CFC's seem to take the brunt of the blame, and should these be the cause of the late twentieth-century warming, the problem would be relatively simple to correct.

CFCs

Many sources confirm that CFC's contributions to global warming are at a small level relative to CO_2. Probably most scientists would concur with this position, and it is not inherently likely to be untrue. However, some of these same sources have used certain atmospheric studies – that include the effects of CFC's – to mask the lack of empirical evidence for observed CO_2 warming since 1970, thus confusing the issue. Although evidence of slight warming from these gases has been easier to implicate, one cannot have it both ways.

Qualified Critics

Since we know there are many who disagree with conventional AGW theories, we should know a little of such perspectives from a few of those who have, at least, significant scientific credentials.

Richard Lindzen, of the Massachusetts Institute of Technology, was intimately involved in the creation of the 2001 IPCC Report and Working Group 1, and a contributor to the 1995 IPCC Second Assessment; he is one of an elite group of leading internationally renowned climatologists. He is widely respected, even among those who take diametrically opposing positions, though both Hansen and Lindzen have been among the most savagely attacked by some of their opposite counterparts.

Lindzen's perspectives perhaps can best be summed up by his likening of the constant changes made to climate models to a classroom exam where students know the answers but arrived at them by using all the wrong methods. He famously outlined a result of technical statements emanating from the scientific community that are beyond the capability of the media to interpret correctly. Restated in alarmist terms, politicians respond by awarding increased funding for research!

He has maintained a position that all dire projections of warming include assumptions that added feedback from water vapor and clouds will greatly multiply this effect, believing that undetermined feedback from water vapor damages the credibility of climate models based on such statistics as provided by GISS. Lindzen repeatedly has claimed that such models have been misrepresented and manipulated through the conduit of the media, and also has criticized the IPCC Summary for Policymakers for not accurately reflecting the report's findings, as previously referenced in the preface [13]. Lindzen has stated for the record that doubling present levels of CO_2 alone would produce a further warming of just 1°C. This is again much in line with T. J. Nelson's calculations (see Chap. 2).

Climatologists Roy Spencer and William Braswell also have weighed in, authoring several papers [14]. It is clear that they consider that models of feedback effects have been "misdiagnosed" and that current observations are "biased in the positive direction." Note that significantly, and in light of earlier remarks, they did not dismiss the effect of greenhouse gases but consider them to have been overstated. Thus, they appear to be looking for truth somewhere in the middle. None of these findings necessarily should be taken to mean the authors disagree that some amount of manmade greenhouse effect could not be in play.

Differences in interpretation of the data can also be tied to other factors. Nils-Axel Mörner, a leading figure in sea level studies, if a highly controversial one, was the former head of paleogeophysics and geodynamics at Stockholm University. He has presented numerous papers, many appearing in *Global Planetary Change*, an important journal. Because astronomical climate research has involved the effects of the Sun on Earth's ocean masses, we can expect that any warming detected within them will be tied to the Sun, no matter how indirectly.

An outspoken critic of anthropogenic warming, Mörner remarked of the IPCC Third Assessment Report (TAR), "It seems that the authors involved in this chapter were chosen not because of their deep knowledge in the subject, but rather because they should say what the climate model had predicted." A gruff and short commentary, with a highly amusing graphic, may be accessed online [15].

Former *Apollo 17* astronaut and now a U.S. senator, Harrison H. Schmitt has tied the differences to outside influences, roundly condemning the National Academy of Sciences for their policies of "promoting a federal mandate based on flawed as well as selective science" within the American Association for the Advancement of Science. Of the latter he expressed distaste that it "could not even bring itself to require consideration of books dissenting from the 'consensus' that current climate change is human caused." More serious still are Schmitt's comments that "if grant applications from the researchers involved do not propose to show the effects of humans on climate, their proposals risk not being funded."

Whether this view represents reality depends on whom one listens to, but it has been widely propagated. One difference, however, between this senator and most of the remainder of his colleagues is that Schmitt does have a real scientific and astronomical background.

Strident tones against AGW (anthropogenic global warming) theories have also emerged from other, sometimes unlikely, places [16]. Taking a decidedly purist approach using mathematics and statistics, along with data from NASA/GISS, economists Michael

Beenstock and Yaniv Reingewertz claimed that the anthropogenic theory is 'spurious,' that the contributions of greenhouse gases are temporary rather than permanent, and that solar irradiance is indeed the driver of climate change. The final opinion was that although the effects of increased carbon emissions were real, they would be only temporary, and likely to be reversed in the near future. It seems that this final position has its feet planted firmly on both sides of the argument!

And Then...

Perhaps the most telling effort to find common ground among all came as recently as 2011 from no less a figure than James Hansen who, in a paper with several co-authors, made some interesting – and indeed rather startling – adjustments to his long-held positions [17]. This is a critically significant paper, as the authors attempted to juggle contradictory evidence, some of it of their own making.

An extraordinary concession made early on in the paper was that effects due to anthropogenic aerosols are virtually unmeasured, along with their influence on cloud formation, having probably been a greater negative forcing than had been assumed. Hansen et al. then continued along some climate-hydrological lines not too far removed from those analyzed by Trenberth (see Chap. 8). The conclusion was that climate models failed to project accurately the absorption of projected greenhouse heat into the oceans, with rates of absorption since 2003 *far less* than had been expected. Another stunning concession was the conclusion that the models must have employed a net climate forcing greater than had taken place in the real world of climate forcing, and that the slow climate response function of the oceans had allowed climate models to compute an unrealistically large net climate forcing.

However, perhaps even more startling were remarks within the summation that since projections of net anthropogenic climate forcing had been unrealistically large, it would signal something of a white flag and could only lead skeptics once again to call into question the viability of models as a whole. Although Hansen lauded the success and effectiveness of the GISS model 1E-R in projecting sea surface warming, its findings had overstated warming of the deep oceans. It had further been estimated that aerosols acted as a cooling agent twice as efficiently as had been allowed.

The mindset of many scientists such as Hansen, who support the IPCC position, has long been that CO_2 increases inevitably lead to continuing increases in temperature, and thus the only explanation they could provide for inaccurate forecasts is that they have not correctly counted the aerosol factor. However, if once again we return to T. J. Nelson's paper (Chap. 2), a more realistic explanation might be that the existing CO_2 is already fully heat-saturated. Additionally, the researchers referenced that the rate of sea level rise had slowed by an estimated 0.6 mm/year, according to ARGO program.

However, the paper did not stop there. In perhaps the most astounding remark, and delivered almost off-handedly, the authors stated that the biggest uncertainty in assessing the fast-feedback effects of increased greenhouse gases was "actual global temperature change."

A stunning conclusion if ever there were one. Accordingly, it was stated that no climate model would ever be able to compete with empirically derived data. Stunning, again. Among further admissions were that previous studies (i.e., Hansen et al. 1984) treated ancient aerosol changes incorrectly as a forcing. In fact, they concluded anthropogenic aerosols were probably a far greater negative forcing than had been assumed, because they could not otherwise explain the slow rate of warming relative to their calculated energy imbalance.

The Effects of Aerosols

Many have claimed that aerosols have been masking global warming, which otherwise would be worse. In light of recent positions taken by leading AGW proponents, and having made such a bold pronouncement before they had evidence, the convenient carelessness of this position can easily be seen, especially since it has been stated in a vacuum. The role of aerosols is still unclear at this time.

Hansen thus revealed some truly remarkable qualities that his critics had always been unwilling to acknowledge or similarly demonstrate: that he remains ready and willing to adjust his conclusions, in an open manner, when evidence emerges that past conclusions have been flawed. This is quite contrary to the positions taken by many figures on both sides in many other instances. Kudos to Hansen. Perhaps there is a way out after all.

Shark-Infested Waters: Surfing for Insights

The example above represents a well-justified rationale. In contrast, a 2007 article from the *National Post* about climate change occurring in other planetary members of the Solar System contained virtually every possible blanket cliché [18]. Relegating some possibly interesting research into the realm of pop science by attributing – without a shred of evidence – that, somehow, the Sun *had* to be responsible for every observation of warming within the Solar System, they fell into their own trap. It is interesting to note that we seldom see this kind of populist, pseudo-science from those supporting the IPCC position, an observation that should not be lost on those taking the opposite stance.

Other official-sounding organizations have taken just as simplistic blanket positions, seldom with much to back them up. The Capital Research Center stated that the Union of Concerned Scientists had waged a "jihad against climate skeptics." Perhaps the worst effect of these is when they link climate science and politics together. From the wealth of materials on the Internet about climate change, independent websites with grandiose, less-than-concealed names and headings are likely to be suspect sources. These sources should not be confused with well-written peer-reviewed research, even though one should not conclude that all independent commentary is necessarily of little or no relevance; we just need to be extremely careful.

However, we can also find instances also of less-than-model practices even from among other government agencies. One closely

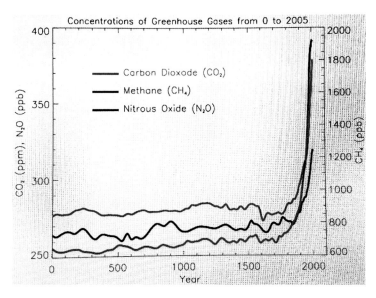

FIG. 9.2 Carbon dioxide concentrations over the past 2,000 years (From the NOAA website)

affiliated with NASA, the National Oceanic and Atmospheric Administration (NOAA), features on its website at least one graph (Fig. 9.2) that uses a base point *other than zero*, resulting in some startling results [19].

The NOAA graph for atmospheric CO_2 increases registers the starting point – 280 ppmv (that's right, 280 parts per million) – at the bottom of the graph, showing it soaring dramatically to the top of the chart with the most recent reading of 388 ppmv. The conclusion likely to be drawn by the uninitiated is that CO_2 has taken over the entire atmosphere! This from a governmental agency, no less. Although many examples from private sites may be just as bad, at least they do not represent public taxpayer-funded official agencies.

Just as appalling, in this writer's view, is the 'fact sheet' that appears with this grossly misleading graph. From the outset, the page reveals a predetermined agenda with stated certainties about anthropogenic warming – remember, even the IPCC itself uses only terms such as "likelihood," or "high level of confidence."

Manmade Global Warming

Many pro-AGW websites, such as the very official sounding, *Environmental Defense Fund*, and the *NRDC*, the *National Resources Defense Council*, state unequivocally amongst their lists of 'facts' that global warming is not a natural process, and that humans and CO_2 are causing it [20]. The NRDC repeats a false position that CO_2 is a heat "trapping" pollutant, while promoting an agenda based on statistics about new employment in clean, green industries. However, these usually include frequently including long-existing employment in trash and waste disposal, hydro-electric power and so forth.

Even the most ardent AGW climate scientists have shied away from stating for the record such simplistic 'black-and-white' positions, and stated in the absence of documentation. *Instead, they usually assign 'levels of confidence' to each aspect of existing research.*

The 31,000

We cannot leave this part of the discussion without briefly touching upon a petition purportedly signed by 31,000 U.S. scientists who have challenged the consensus view. Such dissent, if accurately representing what it appears to, illustrates vividly the wide disparity of interpretations of the same data. Many distinguished names, astronomers and space scientists among them, appear on these rolls, including Fred Singer, Edward Teller, Reid Bryson, William Gray and William Happer – assuming they are one and the same as the famous scientists of those names. One private website claimed that the 31,000 scientists represent only 0.1% of U.S. scientists who hold a BS degree or higher. If 31,000 signatories represent 0.1% of U.S. scientists, this would translate to 31 million, or *10% of the population!* Regardless, of the portion who are connected to astronomical and space sciences, this would translate to an extremely unrealistic number.

The 31,000 signatories in question appear in the Global Warming Petition Project, the creation of noted physicist Frederick Seitz (1911–2008) [21]. Regardless, it was not the first such petition, but it is by far the largest. Regretfully, none of the names on the list state their affiliations, credentials, or published papers, etc., which is its greatest – and some would say fatal – weakness. Although many critics have tried to discredit this source because not every name appearing on the list may be legitimate, we ought not assume that *none* of them are. It is equally interesting to note the many well-known figures who might be considered critics are among those names that do not appear. All in all, it is perhaps unfortunate that the petition was not better devised, more that it is impossible to verify what is presented.

However, the spectre of human-induced (anthropogenic) carbon dioxide levels remains pivotal to the discussion, and the most polarizing part of it. The 'establishment' has not done a good job in providing clarity beyond blanket positions of absolutes while allowing for little curiosity. Worse is when it appears to be outright contradictory. That perception has allowed some to call *all* the field of climate science into question.

References

1. Hansen J, Soto M, Khareshka P, von Schuckmann K (2011) 'Earth's energy imbalance and implication. NASA Goddard Institute for Space Studies (GISS). Atoms Chem Phys 11:13421–13499
2. Hansen J, Johnson D, Lacis A, Lebedeff S, Lee P, Rind D, Russell G (1981) Climate impact of increasing carbon dioxide. Science 213(4511):957–966
3. Hansen J, Lebedeff S (1987) Global trends of measured surface air temperature. J Astrophys Res 92(D11):13345–13372
4. Hansen J, Ruedy R, Sato M, Lo K (2011) 2010 global surface temperature change. Rev Geophys 48:RG4004
5. Brohan P, Kennedy JJ, Harris I, Tett SFB, Jones PD (2006) Uncertainty estimates in regional and global observed temperatures: a new data set from 1850. J Geogr Res 111(D12):D12106
6. Hansen J, Ruedy R, Glascoe J, Sato M (1999) GISS analysis of surface temperature change of 1999. Goddard Institute for Space Studies (GISS). J Geophys Res 104:30997–31022

7. Hansen J, Ruedy R, Sato M, Imhoff M, Lawrence W, Easterling D, Peterson T, Karl T (2001) A closer look at United States and global surface temperature change. Goddard Institute for Space Studies (GISS). J Geophys Res 106:23947–23963

8. Cook J (2010) Empirical evidence that humans are causing global warming. Skept Sci. 2010 http://www.skepticalscience.com/empirical-evidence-for-global-warming.htm

9. Harries JE, Brindley HE, Sagoo PJ, Bantges RJ (2001) Increases in greenhouse forcing inferred from the outgoing longwave radiation spectra of the Earth in 1970 and 1997. Nature 410(6826):355–357; Griggs JA, Harries JE (2004) Comparisons of spectrally resolved outgoing longwave data between 1970 and present. Proc SPIE 5543:164

10. Qing-Bin Lu (2010) What is the major culprit for global warming: CFC's or CO_2? J Cosmol 8

11. Wang K, Liang S (2009) Global atmospheric downward longwave radiation over land surface under all-sky conditions from 1973 to 2008. J Atmos Phys 114:D19101

12. Evans WFJ, Puckrin E (2006) Measurements of the radiative surface forcing of climate. In: 18th conference on climate variability and change 1–7

13. Lindzen RS (2005) Understanding common climate claims. Draft paper for 2005 Erice meeting of the world federation of scientists on global emergencies. eproceedings.worldscinet.com/9789812773890/9789812773890_0016.html, pp. 189–210

14. Spencer RW, Braswell WD (2008) Potential biases in feedback diagnosis from observational data: a simple model demonstration. J Clim 21(21):5624–5628; Spencer RW, Braswell WD (2010) On the diagnosis of radiative feedbacks in the presence of unknown radiative forcing. J Geophys Res 115:D16109; Spencer RW, Braswell WD, Hall C (2011) On the misdiagnosis of surface temperature feedbacks from variations in Earth's radiant energy balance. Remote Sens 3(8):1603–1613

15. Mörner N-A (2004) IPCC again. INQUA commission on sea level changes and coastal evalution, 25 July 2004. http://www.pog.su.se/sea/HP-14.+IPCC-3.pdf

16. Beenstock M, Reingewertz Y (2010) Polynomial cointegration tests of the anthropogenic theory of global warming. Department of Economics, Hebrew University, Mount Scopus

17. Hansen J, Sato M, Kharecha P, von Schuckmann K (2011) Earth's energy imbalance and implication. Atmos Chem Phys 11:13421–13449

18. Gunter L (2007) Bright sun, warm Earth. Coincidence? National Post, 12 March 2007. http://www.nationalpost.com/story.html?id=551bfe58-882f4889-ab76-5cel02dced7

19. How do human activities contribute to climate change and how do they compare with natural influences? NOAA. oceanservice.noaa. gov/education/pd/climate/factsheets/howhuman.pdf

20. Global warming myths and facts. Environmental Defense Fund. http://www.fightglobalwarming.com/page.cfm?tagID=274; National Resources Defense Council. NRDC. www.nrdc.org

21. Global warming petition project. http://www.petitionproject.org/

10. Global Warming on Other Worlds

We have all heard it: global warming is not exclusive to Planet Earth [1]. A number of other members of the Solar System may be experiencing their own forms of climate change. However, before succumbing to the exciting lure that there is, or is not, any relevance to what may be taking place on other planets and what is occurring on Earth, we should look to see if there are actually any parallels at all. Because not everyone has been persuaded by the supposed climate links that have been promoted, we may have also heard absolute denials of the possibility of extraterrestrial warming. Other researchers have taken the position that these perceived changes have no connection with the Sun at all, or to Earth's recent warming (Fig. 10.1).

However, it is certainly easy to understand why some have jumped to the conclusion that if there has been an increase in solar activity, it is only natural to expect that there should be some repercussions in places other than on Earth – assuming we accept even the latter premise.

Although these theories and sentiments are certainly worth looking into, such simple reference points do not necessarily equate with similar phenomena in other places. Climate change in other places does not necessarily follow the same 'rules' as on Earth, for a number of reasons. Whatever may or may not be taking place within the greater Solar System, at present it remains problematic to be definitive about any of it. It is still harder to tie any part of those various observations to any particular factor, for a number of reasons:

- Earth exists within a complex environment, containing almost countless components that react singly, or combined interactively, together. Its highly active, turbulent atmosphere, magnetosphere, variety of landmasses, abundant surface water, deep

A. Cooke, *Astronomy and the Climate Crisis*,
Astronomers' Universe, DOI 10.1007/978-1-4614-4608-8_10,
© Springer Science+Business Media New York 2012

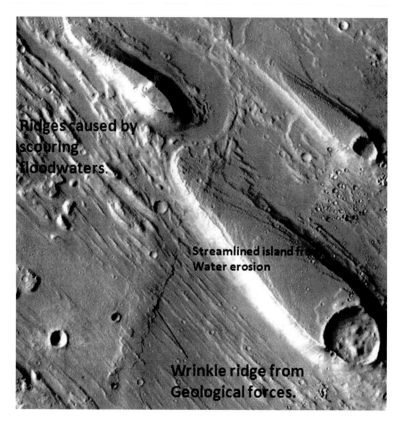

Fɪɢ. **10.1** Evidence of once-flowing water and erosion on Mars (Ares Vallis) (Image courtesy of NASA)

and shallow ocean currents, temperatures, significant atmospheric greenhouse and aerosol conditions, with extremely complex feedback processes, makes it unique in the Solar System. On Planet Earth, many additional influences may apply, well beyond the obvious. It is quite possible not all have yet been established.

- Earth's orbit is also unique; its proximity to the Sun makes it not only a prime candidate for the development of life forms but for multiple strong interactions with the Sun, the central source of all radiant energy. Other planets with more elliptical orbits and far longer orbital periods actually may be moving further from the Sun at this time. Pluto's wildly eccentric orbit even crosses that of Neptune's at times. For other planets, this factor alone could easily cancel completely any potential increased warming from the Sun. However, Earth's less elliptical and smaller orbit ensures more consistent and greater solar exposure.

- Earth has an axial tilt and precessional wobble that is also unique to it. These have significant consequences. Whereas other planets and their satellites also have similar characteristics, these are substantially different overall. Thus, they are affected accordingly at any moment of time, their situation often being totally opposite. Equivalents of the Milankovic cycles on Earth (see Chap. 11) therefore cannot be applied with the same blanket values, and maybe hardly at all.

- The four largest planets are differently structured from Earth, being primarily gaseous in nature. Presumably with no hard or mountainous surfaces below, this affects their response to solar activity. Additionally, their atmospheres have entirely different compositions from Earth's, and their substantially different temperatures greatly affect their behavior and responses to solar input.

- Differences in the magnetospheres of different planets also play a substantial role in climate issues. The interplanetary magnetic field (the Sun's), or those of individual planetary magnetospheres, both have also been implicated in the deflection of cosmic rays. Although deflection of the solar wind is part of the whole, and not directly tied to temperature, its interactions with other related planetary components do indeed affect environments, and ultimately influence climate.

- Because the lengths of planetary years and orbits also vary substantially (Pluto's being the longest at some 248 Earth years), entirely different potential tidal forces of their alignments with other Solar System members have to be considered (see Chap. 7). This could either reduce or enhance possible solar interactions.

The Sun, as ever, remains at the center of all controversies, and no less so with climate issues concerning other members of the Solar System. Despite the case that has been made for solar irradiance having increased through the turn of the twenty-first century, (as laid out by the ACRIM team and others), the controversy has never gone away. The PMOD team has not given ground in the argument and remains closely tied to the IPCC position (Chap. 5). Old arguments thus die hard about the Sun's recent output, and continue to be raised in the most public of forums. If the

Sun has been cooling in recent years, we should not expect to be witnessing solar-induced warming on other worlds, especially since in the most quoted instances the warming and cooling effects ought to be more or less instant in the instances in question.

An article that appeared in *New Scientist* [2], 'Climate Myths: Mars and Pluto are warming too,' typified another all too common type of argument. By again selectively choosing only the much-challenged solar outputs (the alternative PMOD analysis), the article stated without the slightest hesitation that the Sun's output had *not* increased since 1978. Thus, even *if* the wrong argument was used to refute another, the implication was that no warming could have taken place anywhere because the Sun's output had declined.

Connecting the Unconnectable

It might be instructive to spend a little time with the website, globalwarminglies.com. Here you will find all manner of arbitrary connections made between what is happening on Earth and what has been observed on other planets. The comparisons made are patently shallow – ultimately absurd, as attempts are made to correlate what cannot be correlated in any possible way.

So where do we begin? First, we should acknowledge that there is no doubt that changes have been observed in the surface conditions of other members of the Solar System. However, before we jump on the bandwagon of believing this is just another manifestation of what we are experiencing on Earth, we should examine each case individually. In doing so, we might be able to even test some climate theories about Earth's own situation. Additionally, some events might only be occurrences that went undetected before.

Venus

Although this planet is not among those that have been thought to be currently experiencing climate change, looking at the processes surrounding its dense carbon dioxide laden atmosphere is certainly instructive. Many researchers have looked to it for parallels and findings on Earth's greenhouse conditions, because this planet perhaps represents the ultimate extreme case of CO_2 greenhouse warming. However, the warming definitely is not manmade! (Fig. 10.2)

Direct comparisons with Earth's present situation are far from productive, since the Venusian atmosphere is virtually all CO_2 (96.5% CO_2, 3.5% N_2), compared to the puny proportion of CO_2 in Earth's atmosphere at 0.0388%. There is really no way to compare the two planets beyond their similarities in size, since even these most basic of circumstances could hardly differ more. The atmosphere is also a hundred times as dense. Fortunately, there are other facts about greenhouse conditions we can learn from Venus.

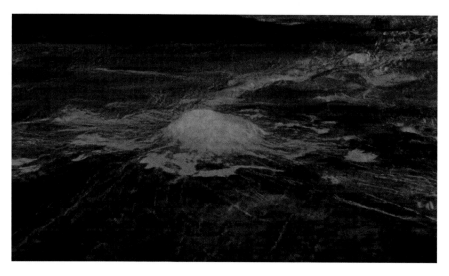

FIG. 10.2 The surface of Venus in the vicinity of a volcano (infrared image) (Image courtesy of NASA/JPL-Caltech/ESA)

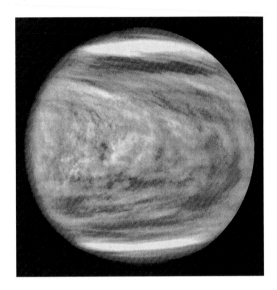

FIG. **10.3** The dense atmosphere of Venus (Pioneer) (Image courtesy of NASA)

The extreme opacity of its atmosphere is due to clouds of sulfuric acid droplets and sulfur dioxide. The temperature at the surface is a scalding 737 K, which registers almost 500 K higher than that on Earth. Some of this high temperature is determined by its closer proximity to the Sun, but if we remember that almost all of the atmosphere of Venus is greenhouse gas, versus, at most 2% of that of Earth (and only if we include Earth's highly variable water vapor, the largest greenhouse component), a better perspective emerges. In addition, Earth's atmosphere is also partly warmed by trace greenhouse gases other than CO_2. Once we include the multiple feedback from all of them as well, we can see how greatly the environment of Venus differs (Fig. 10.3).

A recent article in *Nature GeoScience*, Nov. 16, 2010 [3], reported the confirmation of significant sulfur dioxide in Venus's atmosphere. Possible cooling effects of the transformation into sulfuric acid high in the atmosphere were also discussed, although this was not as groundbreaking as it might seem. However, in the 1970s, sophisticated analyses by ground-based instruments had already enabled the deduction of the presence of such aerosols high in the Venusian atmosphere. At the time no less a figure than James Hansen had conducted groundbreaking research into such potential cooling by these compounds. He had concluded that any possible reflective properties that the clouds might have contributed against incoming solar radiation were far outweighed by the warming caused by heat reflection back to the surface. Significantly, it was largely through Hansen's efforts that the effects of such sulfur particles in Earth's atmosphere would become appreciated and understood for their role in climate.

Because Earthly aerosols emitted by volcanic eruptions are largely comprised of sulfur compounds, too, the possibility of using them to reverse warming trends has been contemplated. It has been even the subject of actual programs and proposed experiments (see Chap. 2). Regardless, It would seem that further thoughts of experimenting with global cooling processes from artificially adding these compounds into Earth's atmosphere are presently not in NASA's cards. Regardless, others have continued to contemplate, theoretically at least, the possibilities, such as chemistry Nobel Laureate Paul Crutzen.

Paul Crutzen has proposed a method of artificially cooling the global climate by releasing particles of sulfur into the upper atmosphere to reflect incoming sunlight and heat back to space. Some leading scientists have taken the controversial proposal seriously because Crutzen has a substantial and proven track record in atmospheric research.

A diametrically opposing view [4] is that of Nasif S. Nahle, a biologist who has taken an intense interest in climate studies and has self-published his findings online. Although all of his climate work appears to be independent of any publisher or peer reviewing process, it does not appear to be in the realm of pop science. However, it must be appreciated from the start that Nahle is also very independent in the conclusions he reaches. Although he leads us with great care through the entire system of his deductions, the result seems completely incompatible with the research of other scientists in the mainstream – or perhaps anywhere! His conclusions are that the amount of CO_2 greenhouse warming on Venus would total only 7.36 K, and thus, CO_2 super-heating of that planet – and CO_2 greenhouse warming of Earth – is a myth!

Naturally, such estimates are far out of step with any of those we discussed in Chap. 2. This is especially the case here, since it is considered that less solar energy reaches the Venusian surface than it does Earth, due to the great reflectivity of the cloud cover. There does not appear to be any parallel study, so Nahle truly is out on his own.

The Mercury Factor

Further difficulties in reconciling Nahle's conclusions by any other scientific logic continue, since Venus's temperatures average 64°C more than those of Mercury, an airless planet that is almost twice as close to the Sun. However, his position seems set in stone, and his denial of what has been observed and extensively analyzed apparently is irreconcilable.

Nahle's explanation? Solar plasma particles have easy access to the planet through a weak Venusian magnetosphere. The circumstances themselves are not in dispute. The lack of a significant planetary magnetosphere was first observed by *Mariner 1* in 1962 and confirmed by *Mariner 5* as well as subsequent missions. Instead, what limited magnetic field as Venus does have appears to be 'induced,' as the bow shock formed when its ionosphere collides with the larger interplanetary magnetic field, compressing the wave front as it comes into contact with it [5]. Apparently, this has the double benefit of causing some ionization in this part of the atmosphere, assisting in deflecting solar winds that would otherwise erode it.

But could this have produced such an outcome? Data gathered in 2008 by the NASA Messenger mission indicated that although Mercury's magnetic field was strong enough to deflect the most severe of solar wind [6], it is far less than that of Earth. Measurements are generally in the 1% range, so essentially it is not very different from that of Venus. The fact that the solar wind is not generally considered to be part of the process of warming of an environment such as this also stands out. Thus Nahle's scenario, while providing some highly interesting speculation, does not seem sufficiently plausible to have produced such a dramatic outcome (Fig. 10.4).

FIG. **10.4** Mercury (Image courtesy of NASA/JPL (Messenger))

Accepted theories continue to point to an extreme Venusian greenhouse condition. However, there are other independent papers that have come to similar conclusions to those of Nahle, if arrived at by different processes. Another website, operated by J. F. Anthoni, concludes that Venus's temperature is primarily caused by convection and not greenhouse effects [7]. In his online article, Anthoni cites two other independent studies that reject greenhouse warming entirely [8], which would have repercussions for Earth's climate projections. Should any of these be taken seriously? The absence of similar analysis in peer-reviewed research would seem to speak loudly. However, all of this makes for a compelling discussion, and perhaps even for a modicum of further investigation into some of these peripheral concepts.

Martian Water

Mars, the subject of probably the most intense extraterrestrial controversies, unsurprisingly can count climate change among them. In recent years, with each new discovery of physical activity on the Red Planet, numerous commentators have been swift to

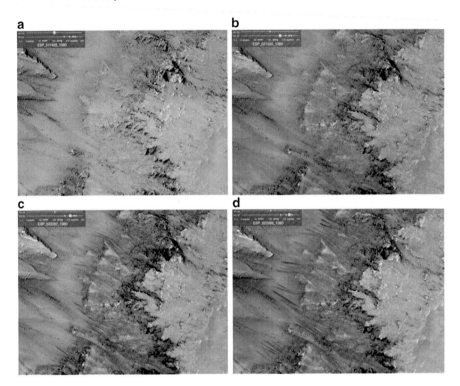

FIG. **10.5** Water on Mars (**a**), beginning of flow, through (**d**) maximum observed flow (Images courtesy of NASA)

pronounce parallels to Earth's own changing climate [9]. Sometimes even serious astronomers have joined in. However, all may not be quite what it may seem at first blush.

An excited announcement in 2011 about recent observations by NASA's Mars Reconnaissance Orbiter revealed what seems likely to be water released from buried ice stores, flowing down and saturating the soil of some crater slopes in ever-darkening streaks (Fig. 10.5a–d). This is an environment that was not supposed to be conducive to free form *flowing* liquid water, one in which it was thought not so long ago would cause evaporation immediately upon ice liquefying – at almost all altitudes and all possible Martian temperatures. As a result, many individuals were swift to proclaim this example of possible Martian water flows as proof of global warming on Mars. In the absence of human-induced warming on Mars, others immediately cited this as being the result of the same natural warming mechanisms that were responsible for changes being experienced in Earth's environment.

However, again before jumping to hasty conclusions, we should realize that Mars's environment and circumstances in the world of planetary science are totally different from our own. Aside from a highly elliptical orbit, with a 19% variance from perihelion to aphelion (due to the lack of a comparable influence from a satellite such as our own sizeable Moon), Mars is not subject to the same constraints as Earth with regard to its axial tilt. Although currently its tilt is not dissimilar to Earth's (at 25° vs. Earth at 23.5°), it experiences wild swings of as much as an estimated 40–60° over time.

Mars's perihelion also occurs at a different time each full orbit. Similarly, comparable to Earth's own Milankovic cycles are Martian astronomical cycles (see Chap. 11), but they are totally different, sharing no common traits. Additionally, Martian north to south precession is almost seven times longer than our own. Consequently, we cannot overlook these most basic comparisons, as the two planets thus have entirely different natural forces acting upon them. This is even before we get to physical conditions themselves on the Martian surface or in its atmosphere.

Mars has an extremely tenuous atmosphere, with a surface average pressure of just 1/169 that of Earth's. Even though it consists of 95% CO_2 (plus multiple trace gases), it could hardly be considered a candidate for significant greenhouse warming. Indeed, it seems that the thin, dusty lower atmosphere achieves the opposite effect, actually cooling the surface while the upper atmosphere experiences searing temperatures. However, it is generally thought that at one time Mars would have had a much greater atmospheric presence, since it is clear that there were large flows of water on its surface millions of years ago (Fig. 10.1). Thus, the Red Planet has undergone greater climate change than most of us could even imagine.

Ingo H. Leubner, who we encountered earlier with regard to his theories of solar radius change and the consequences for the Solar System, also weighed in with his research and some specifics for Mars [10]. Using the same methodology as before, he constructed a model for the evolution of the entire Solar System, estimating that water on Mars froze into ice deposits between three and four billion years ago, when Mars's average orbit increased by 30 million km to the present 228 km. Leubner has continued to explore his theories about the orbital dynamics of the Solar System; his studies and findings may be referenced at the American Geophysical Union website.

The lack of Martian magnetosphere has been frequently cited as the main reason for its thin atmosphere, solar winds eventually having destroyed it. However, this argument does not hold when considering the intensely thick atmosphere of similarly magnetosphere-deprived Venus. Significantly, where a strong magnetosphere is present, it cannot be crossed by hydrogen ions, or, in combination with water vapor in the upper atmosphere as H_3O^+ (hydronium ions), in this case the product of cosmic radiation. It is therefore considered that the Mars of ancient times was abundant in flowing water, even oceans, but because of its weak magnetosphere it gradually evaporated into the atmosphere and was lost to space. Apparently Mars also has been unable to hold on to most other atmospheric components, either, due to the power of the solar wind and the planet's feeble gravity.

Water, Atmosphere and Gravity on Other Planets

A key factor on climate within the Solar System, and especially for any possibility of harboring life, is water. In the case of Mars, it seems it has been slowly eroded from the atmosphere due to the absence of a magnetic field and weaker gravity (it has just 38% of Earth's). However, Venus may never have had any water at any time, despite its much stronger gravity being almost the same as Earth's (90%).

Interestingly, trace amounts of methane and water vapor, both gases with a short shelf life in the absence of a planetary magnetic field, have been detected in the Martian atmosphere by various space-based programs since 2003. This points to the possibility of a recent past (or even present) biological origin, and as such, is at the top of ongoing research programs focused on Mars. However, it is much too soon to draw any conclusions, although the prospect is tantalizing and intimately connected to Earthbound climate science, both of those gases being greenhouse gases.

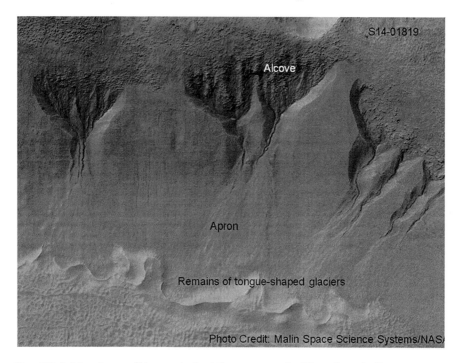

F IG . **10.6** Martian gullies and glacial remnants in Phaethontis (Image courtesy of NASA)

Meanwhile, the presence of gullies on crater and canyon walls has been known since 2000, when NASA excitedly announced the dramatic news of the discoveries (Fig. 10.6).

It was concluded early on that there were limited possibilities for the formation of such features, other than by a flow of some type. These were also generally considered likely to be the relatively recent product of 'newly' released water (within the last million years or so). It was reasoned at the time that their existence appeared to stem from the melting of Martian snow – not underground springs but pressure-induced flows (and certainly not rainstorms!).

For any flow to occur, slopes would have to face away from direct sunlight and its warming effects, and hence, it had been thought that such happenings could only take place at the poles, seemingly eliminating the equatorial regions. However, flows had been spotted *far* from the polar regions, even in those very equatorial regions. For some pop science followers, this could *only* mean

Sun-induced Martian global warming! However, a fundamental difference between Earth and Mars was perfectly illustrated instead, because Mars' peculiar variations of tilt enables shaded slopes to occur almost anywhere on the planet. They can also face either direction, depending on the circumstances of tilt and Mars's position in orbit at the time.

Further discoveries of newly formed Martian gullies and rolling boulders were announced in 2005 [10], indicating that the Martian topography was far from inactive. Once again, could water be implicated? An answer eventually came that was not perhaps what most researchers were hoping for: a new study in 2010 [11] settled on the thawing of frozen carbon dioxide, rather than water, as being the most likely explanation for the phenomenon. For most planetary scientists, this must have come as a big disappointment.

As fortune would have it, those scientists did not have to wait too long, with the surprise announcement in 2011 of a stronger case for water flows being responsible for the gullies after all. Indeed, it appeared, amazingly, that those pictures snapped by the HiRISE camera aboard the Mars Reconnaissance Orbiter were of *present-day* water flows (Fig. 10.5a, b). Dark projections, growing in length with time, down some crater slopes reopened the case for free-flowing water of some kind on the Martian surface. Most exciting was that it had been happening right under the researchers' noses.

While being careful not to claim a conclusive finding, NASA was upbeat and confident of the analysis of the observations, stating that the potential water flow was likely briny in nature, and that it had been observed over successive orbiter laps around the planet. Because briny solutions are more able to become liquid at low Martian temperatures, rather than remaining frozen, it seemed to be the only way that such events could have taken place. Questions remaining included whether such occurrences were unusual (and possibly indicative, therefore, of a warming trend), or if they have occurred routinely in the past.

Melting Ice Caps?

However, perhaps the single-most significant suggestion of possible Martian global warming came in 2005, when it was announced that NASA's Global Surveyor and Odyssey reconnaissance satellite

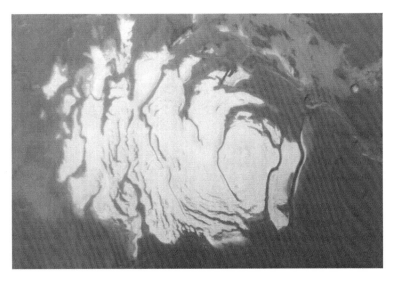

Fɪɢ. **10.7** South polar cap, Mars, 2000 (Image courtesy of NASA/JPL/Malin Space Science Systems)

programs had observed a sudden shrinkage of the south polar cap, a phenomenon of carbon dioxide sublimation that had been taking place over three Martian years during the summers, revealed also as what the researchers termed a 'Swiss cheese-like' thinning in pockets of the carbon dioxide upper polar ice coating (Fig. 10.7).

Although it is normal for the polar caps to recede during the summer months, the degree to which the melting had progressed beyond normal seasonal thaw was apparently unprecedented since such detailed imagery became available. It was also noted that low latitude regions had been observed to harbor more water-ice than would be expected [12]. All of this was considered indicative of possible, though far from confirmed, evidence of Martian climatic change (Fig. 10.8).

Although NASA scientists had reported a possible warming of Mars as responsible, they took care not to link it to what has been experienced on Earth. It was considered most likely to be internally generated. However, Habibullo Abdussamatov (head of space research at the Pulkovo Observatory), whose name we have come across earlier in this book, jumped to the startling conclusion that this observation represented further evidence of increasing solar activity [13]! Despite recent satellite evidence that solar

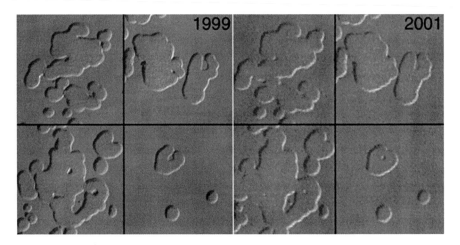

FIG. 10.8 Development of carbon dioxide 'Swiss cheese holes' in Martian southern polar cap (Image courtesy of NASA)

activity had been waning, Abdussamatov, never one shy to take controversial positions, had staked his claim. Basing his studies on solar variations and comparisons of climate patterns on Earth, Abdussamatov's pronouncement about Martian global warming created a stir in the astronomical community, where it had been generally concluded that the cap's change in dimensions was something of a regional anomaly, rather than global in nature, or representative of a fundamental change in climate.

Regardless, despite Abdussamatov's distinguished background, many fellow astronomers were quick to dismiss his hypothesis. They noted that he had not been a supporter of IPCC assessments, especially in respect of the greenhouse effect, which he had long rejected. Certainly, he had reason to look for new evidence to support his beliefs. As for some of Abdussamatov's other views on climate change, he is also among those who consider that Earth is now in a cooling phase, something that logically we might have presumed he would have thought would also be the case on Mars. Less clear is how he equated the multitude of other variables listed at the beginning of this chapter with such simple parallels to what has been happening on Earth.

One could argue quite plausibly that the very large swings in axial tilt may also play a key role in climate fluctuations on Mars.

This does not apply to the present situation, though, because these changes are slow progressing, with no sudden degree of change being possible or having taken place from before observations of the polar cap recession to the present. Despite the similarity of Mars' *present* angle of axial tilt to Earth's, the Red Planet experiences more pronounced seasons and climatic patterns nevertheless, due to its greater orbital eccentricity.

However, simply because solar activity had been declining noticeably at the time the polar cap was observed to be shrinking (and this is not in dispute), a direct solar link does not seem possible from this standpoint alone. Such variations cannot be so immediately and directly tied to climate variations on Earth, though. The lack of a significant Martian atmosphere (and any oceans) means that any changes in the Sun's output will be felt at once on Mars's surface. In comparison, on Earth, the buffering from the Sun's rays through its complex warming and cooling mechanisms may take *years* to be reflected in changes to the climate. An extremely detailed rebuttal and analysis of various solar warming hypotheses on Mars, by Steinn Sigurdsson (Penn. State University), may be found online [14].

Storms on Jupiter

Adding further ammunition to the arsenals of speculation about climate change in other places within the Solar System was an article that appeared in 2004 in *Letters to Nature* [15]. The author, Philip S. Marcus (a professor of fluid dynamics at University of California at Berkeley, and one of the Hubble team), surmised that we might soon be able to observe noticeable changes to the cloud tops of Jupiter due to the disappearance of many anticyclonic vortices (the familiar white oval 'spots'). This would bring about a warming effect of approximately 10 K (Fig. 10.9).

Marcus's theory proposed that anticyclonic vortices were being trapped by Jupiter's equivalent of Rossby waves, largely eliminating the present cyclonic Jovian 'spots,' a process with a lifespan averaging 60 years (another 60-year cycle?). A recurring sequence of events would result in the ongoing creation and destruction of the vortices. Their loss would mean less venting of

Fig. 10.9 Jupiter and spots (Image courtesy of NASA/JPL/University of Arizona)

atmospheric heat, leading to increased global temperatures. In turn, this would create atmospheric instability, ultimately ushering in a new cycle of anticyclone formation, with the familiar reappearance of oval spots. At such point, it would seem that temperatures again would drop to prior levels.

Perhaps the most important point to note is that *at no time did Marcus state that temperatures on Jupiter actually had been observed to rise.* No matter. It was enough for certain zealous climate conspiracy buffs to take the ball and run with it, proclaiming this to be 'another' example of a naturally occurring solar-induced climate change in the Solar System. This was surely not what Marcus had in mind. Many press articles simply misstated the facts, at least one of them even mixing up 10 K with 10°F. Again, such definitive – even dogmatic – declarations of Solar System-wide climate mechanisms are non-sequiturs, all with different causes and effects. In this case, they represent only what was *proposed* as a potential theory of a warming effect! As of the time of this writing, such warming still has not been detected.

Other researchers consider that many of the smaller white cyclonic storm-spots will slowly merge into a larger great storm and disappear. Perhaps there was a link to Marcus's theory after all. There is some evidence of this type of evolutionary process taking place, as illustrated in Fig. 10.10.

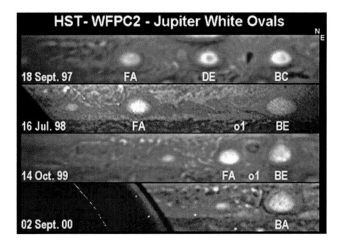

FIG. **10.10** Merging white spots (Image courtesy of NASA)

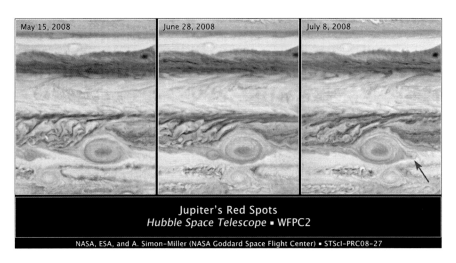

FIG. **10.11** The Great Red Spot, Red Spot Jr., and 'Baby Red Spot' devoured (*arrow at right*)

Meanwhile, the pronounced coloration of the Great Red Spot (GRS) is something of an anomaly; most spots are white. When the three small white oval spots in Fig. 10.10 started to converge on the iconic Great Red Spot's domain in 1999, they eventually merged to form one larger spot. By 2006, this new spot (ultimately termed 'Red Spot Jr.') had survived several close encounters with the GRS and was seen to be taking a similar reddish hue (Fig. 10.11).

We could speculate, of course, that color changes are connected to temperature, with perhaps the strength of the upward rising vortex exposing deeper-lying mixtures of warmer gases, or even the altitudes of the tops of these twisting systems rising above obscuring gas layers. That same year, an article appeared in *Science Daily* [16] that described well all that had been occurring on Jupiter.

Since that time the Great Red Spot itself appeared to have spawned yet another, third smaller red spot (the 'Baby Red Spot'). Scientists speculate that both new spots should be considered possible candidates for consumption by the massive larger spot, perhaps illustrating the mechanism that has sustained it for almost four centuries. Alternatively, the outcome could be yet more reddish spots! Scientists had observed increasing turbulence on both sides of the GRS, along with considerable variance in the hues of the cloud belts in general. Could they possibly be further indications of Marcus's global warming prediction as the atmosphere reacts? However, it is still important to stress that no actual Jovian warming has been detected at this time!

By 2008, the three red spots were still independent until one fateful encounter with the GRS resulted in the demise of the baby spot (Fig. 10.11). The Great Red Spot apparently is capable of eating its young.

All in all, however, it is not possible to say what has been taking place on Jupiter recently, and especially if it is indicative of a warming trend to come. Meanwhile, all claims of Jovian global warming should be taken with maybe more than a pinch of salt.

Triton: Neptune's Warming Satellite?

Rumblings about global warming on Triton also helped to fuel what has become an endlessly compounding gossip mill. However, again, we need to look at exactly what was observed on this distant world before jumping to any foregone conclusion (Fig. 10.12).

As the most distant of the Solar System's eight recognized planets (since Pluto's recent fall from grace), Neptune has 13 known satellites. Of them, Triton is presumed to be a former planetoid similar in structure and size to Pluto, and captured from

Fɪɢ. **10.12** Artist's impression of the surface of Triton and its thin atmosphere (Image courtesy of ESO/L. Calçada)

among the larger objects of the Kuiper Belt in the early days of the Solar System. Its retrograde orbit around Neptune and extreme inclination (presently 129.8° relative to the ecliptic), along with its various physical properties would seem to confirm a less than impeccable planetary satellite origin. Orbiting in lockstep to Neptune's rotation, as does Earth's Moon, it keeps the same face turned to its parent planet at all times, except Triton itself is slowly being drawn inwards in an ultimate path to tidal disintegration.

With a surface coated in frozen nitrogen, water and carbon monoxide, Titan's overall makeup consists of up to 45% water ice, the remainder consisting of rocky compounds. Its atmosphere is extremely tenuous, comprised primarily of nitrogen that has sublimated from the surface, along with other trace gases. These include carbon dioxide and methane, the latter of which has provided food for any amount of wild speculation about potential life forms. The western hemisphere exhibits a curious texture that has been termed 'cantaloupe terrain,' which is believed may have resulted from flooding by the hot processes of geysers or volcanism, or even the separation into uneven layering of different materials of various densities (Fig. 10.13).

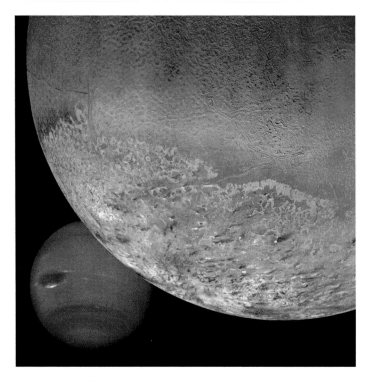

Fɪɢ. **10.13** Triton and Neptune (cantaloupe terrain clearly visible) (Image courtesy of NASA)

In 1998, an edition of *MIT News* [17] featured an article about global warming on Triton. Massachusetts Institute of Technology researcher James Elliot had concluded there had been a noticeable increase in the density of Triton's atmosphere at the expense of the frozen nitrogen on the surface. Utilizing one of the Hubble Space Telescope's fine guide sensors and other sophisticated instruments on Earth during an occultation of a star, the gradual decrease in brightness prior to its being extinguished provided a measure of atmospheric density. This measurement allowed a final calculation of atmospheric pressure. It was thus determined that since last measured in 1989, some 9 years previously, Triton had experienced a substantial increase in atmospheric pressure, corresponding to 1.7°C.

A possible explanation given was that Neptune's location along its 164-year elliptical orbit had brought it close to its summer season, a rare placement that was even closer to the Sun this

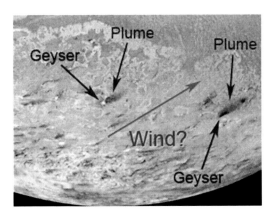

FIG. 10.14 Triton geysers (Image courtesy of NASA)

time than usual. Beyond this possibility, it was thought that variations in the crystalline ice or frozen nitrogen structure might have affected reflectivity. We might even speculate that an unusually active period of geysers or volcanism could have deposited dark ash on the otherwise highly reflective surface. Indeed, active geysers and even hydrocarbon ejecta have been imaged in Triton's atmosphere. Might it be conceivable even that the surface could have picked up some Kuiper Belt dust along its travels? (This last supposition is just the writer's wild speculation ...) (Fig. 10.14).

Triton's normal average temperature of 35.6 K, although not being far from absolute zero, is warm enough that small differences register like large differences would in our own environment. Thus the deduced increase of 1.7 K actually represents an increase of 5%, not an insubstantial amount – enough, in fact. to cause frozen nitrogen to sublimate into a gas, and to make a measurable difference in the density of the atmosphere. Such a large temperature percentage in Earth's climate would be calamity from which all life on Earth probably would perish, apparently comparable to a 22°C warming.

Regardless, in no way does it appear that what has occurred on Triton infers a warming caused by increased solar radiation. Of course, any direct warming effects from a possible increase in solar irradiance of just 0.01% at this enormous distance would likely be very hard to measure. Plus, there is little possibility of indirect solar warming of any kind. More significantly, and perhaps fatally for climate conspiracies, there have been no observations of comparable sudden warming anywhere else in Neptune's own climate, or that of its other 12 satellites.

A Warming Pluto?

It was inevitable that still other examples of out-of-the-ordinary warming phenomena would be found within the Solar System and reported. Interestingly, in 1988 and 2002, the same basic temperature measuring techniques were utilized for Pluto, as had been applied earlier with Triton, except with different instruments located around the world [18]. James Elliot again led the team. Pluto is not considered a particularly likely candidate for reacting to changes in solar irradiance, which also would be miniscule at its great distance, especially at aphelion. More significant to bear in mind is the little world's highly elliptical, lopsided orbit (greater than any planet), which brings it alternately much closer to the Sun than would normally be the case. Additionally, it is placed at an oddly inclined angle to the ecliptic, as well as an extreme angle of axial rotation. All such irregularities eliminate direct comparisons with Earth's situation.

Other possible variations include changes to the surface albedo, perhaps due to some kind of volcanism, but more likely due to variations in frost and methane deposits. A 1988 paper [19] dealt exclusively with this possibility, concluding that the general lack of sunlight, and presumed lack of protection from the solar wind, would cause the frozen surface methane to darken. It is not known if Pluto has a magnetosphere, although it is possible that it does through its barycentric orbit with Charon, its moon. It is not generally thought possible for bodies predominantly made up of icy compounds to have magnetic fields, but examples have been discovered, most notably Ganymede, one of Jupiter's four Galilean satellites.

It is also possible, if it should be that methane darkening is taking place, that the increasingly cold conditions caused when the planetoid becomes more distant from its approach at perihelion would cause gases in the atmosphere to freeze on the surface in the form of frost deposits. It is clear from the images in Fig. 10.15

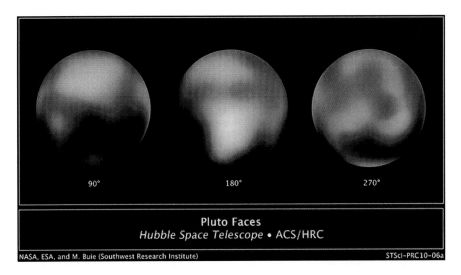

90°   180°   270°

Pluto Faces
Hubble Space Telescope • ACS/HRC

NASA, ESA, and M. Buie (Southwest Research Institute)   STScI-PRC10-06a

Fig. **10.15** Pluto through a 180° rotation

that Pluto's surface albedo is highly varied, both in density and color; the south polar region is decidedly darker than the north, and we may presume that this was determined by the Plutonian pole that was facing the Sun at the time.

Despite Pluto orbiting at a greater average distance from the Sun than Triton (even when Pluto's orbit does cross that of Neptune, it is only by a small amount), the amount of temperature increase (2 K) has apparently been greater than that which occurred on Triton. Pluto's average temperature (44 K) is a little warmer than Triton's, so the ratio of the increase is in some ways compatible for Pluto. Other comparisons are useful in this discussion.

Although Pluto is smaller than Triton, they both have similar sizes (Triton 1,353 km vs. Pluto 1,161 km), similar icy and rocky makeup, with a large presence of frozen surface nitrogen, water and carbon dioxide, and similar densities. Pluto's radical axial tilt to the solar equator (119.6°) and Triton's strong leaning from the ecliptic (129.8°) result in a fairly similar angular relationship to the Sun, atmospheres consisting primarily of nitrogen, along with carbon monoxide and methane, and even relative comparable distances (at perihelion) from the Sun. It has been speculated plausibly that a delayed observable warming following Pluto's closest encounter with the Sun some 13 years earlier might explain the increased temperature (Fig. 10.16).

FIG. **10.16** The surface and atmosphere of Pluto, with its satellite, Charon, low in the sky (Image courtesy of ESO/L. Calçada)

However, the last point about delayed onset warming (LTP again?) is instructive, because it is only at perihelion (when conditions are warm enough to produce it) that Pluto has any measurable atmosphere at all. Pluto last was at perihelion in 1989, and therefore would have had some atmosphere over the period when the observations were made. However, its density is never more than a maximum of about 1/100,000 that of Earth, while that of Triton apparently remains fairly constant at about 1/70,000. MIT researchers therefore concluded when putting everything into context that, despite the similarities of these quite similar worlds, Pluto's atmosphere was more prominently and dynamically impacted than Triton's [20]. It is important to note that Pluto would have been losing its atmosphere during the 14-year period between the observations made, making the increase in temperature all the more striking.

Could all these factors, therefore, possibly imply a common denominator, perhaps even a solar connection to these recently observed temperature increases? In a study from 2005, Henrik Svensmark pointed out that changes observed on other planets could possibly be linked to galactic cosmic rays [21]. He referenced a lessening of Neptune's brightness that seemed to be in step with the solar cycle and changes in brightness (8%) on Titan (a Saturnian satellite). The connection seems plausible, if not proven. We will spend more time with Svensmark's theories in Chap. 12.

It is considered, however, likely that such occurrences are more in the normal range of ongoing variances than exceptional happenings. Because Pluto's surface is also much more highly contrasted than that of Triton, with its extremely dark and bright regions, and therefore entirely different reflective or absorptive properties, comparisons are made harder. Indeed, extensive volcanism has even been proposed as the underlying cause of the recent warming, although there is no evidence of that to date.

Thus, while not quite in the category of comparing 'apples to oranges,' a direct comparison of Triton and Pluto is not likely to be reliable. And although the studies both were conducted by atmospheric analysis, possible connections to a larger external cause are still seen as unlikely. It is also possible that the measurements will later be disproved. Regardless, our knowledge of these distant worlds is so sketchy at best that far more questions remain than answers.

Putting It All Together

Looking at the severe conditions that dominate the environments of the planets, it is hard to not to feel overawed by an awakened appreciation of the little protective bubble that encloses our own Earth as it roams the vastness of space. Looking at the Solar System as a whole, our isolation and the delicate balance of factors we take for granted take on a new meaning and reality. The dependence on our cosmic cocoon to protect, nurture and shelter us is never more starkly apparent, as are the perils that would await us in a moment were we to lose the safe harbor it provides. The Solar System stands in bleak testimony to the fate that stands just a short distance above the atmosphere, were we to find ourselves further or nearer to the Sun, or with any change to the atmosphere great enough to make it instantly toxic or somehow damaged beyond repair. However, perhaps we are living under an assumption that Earth is frailer than it is. As an egocentric species perhaps we take our role in the great order of things too seriously. Perhaps Earth is able to regulate the climate in ways of which we are unaware. Perhaps Earth is the Great Regulator, after all. Only time will tell.

Having examined the most significant of the various Solar System members to which warming has been attributed, it remains problematic to attempt to find a common thread linking any of them, let alone all. Indeed, examples of uneven temperatures exist throughout the Solar System. An example is the 10 K warmer south polar zone on Neptune that will one day shift back and forth to its north pole in a recurring pattern. Overall, it seems not unreasonable to presume that many temperature irregularities, either permanent or temporary will crop up continually within the Solar System, and it is perhaps unwise to read too much into them – at least as far as trying to find a direct tie-in to Earth's present climate situation.

Regardless, this has not stopped large numbers of astronomically unsophisticated individuals from trying. So until there really is any clear evidence of a Solar System-wide warming pattern, it would seem perhaps best to exercise extreme caution in sequestering the jury away to deliberate the case. And however unlikely such a scenario may seem to this writer, it is not the purpose of this book to render a verdict.

References

1. The whole solar system is undergoing global warming, Fragile Earth (2006). http://www.abovetopsecret.com/forum/thread221608/pg1
2. le Page M (2007) Climate myths: Mars and Pluto are warming too. New Sci, 16 May 2007
3. CNRS (2010) 'Sulfur dioxide in Venus' atmosphere could be key to fighting global warming on Earth. Nat GeoScience. http://www.physorg.com.news/2010-11-sulfur-dioxide-venus-atmosphere-key.html
4. Nahle NS (2010) Determination of the effective total emissivity of the carbon dioxide in the Venusian atmosphere, and the mean free path length and crossing lapse (delay) time of photons into the troposphere of Venus. Clim Realists. Online: http://climaterealists.com/index.php?id=6153
5. Luhmann JG, Russell CT (1997) Venus: magnetic field and magnetosphere. In: Shirley JH, Fairbridge RW (eds) Encyclopedia of planetary science. Chapman and Hall, New York

6. Mercury's magnetosphere fends off the solar wind. University of Michigan Public Release, 30 Jan 2008. http://www.eurekalert.org/pub_releases/2008-01/uom-mmf013008.php

7. Anthoni JF (2010) Marble in space; main influences on Earth's temperature. Online: http://www.seafriends.org.nz/issues/global/climate1.htm

8. Duffman HD (2010) No greenhouse effect. Online: http://theend-ofthemytery.blogspot.com/2010/11/venus-no-greenhouse-effect.html; Frizius R (2001) Venus atmosphere temperature and pressure profiles. Online: http://www.datasync.com/~rsf1/vel/1918vpt.html

9. Hinderacker J (2005) Global warming on Mars. PowerLine, 21 Sep 2005. http://www.powerlineblog.com/archives/2005/09/011571.php

10. Leubner IH (2006) Mars orbit and temperature: why and when an early wet mars. American Geophysical Union, Fall Meeting 2004

11. Orbiter's long life helps scientists track changes on Mars. NASA press release, 20 Sep 2005. http://mars.jpl.nasa.gov/mgs/newsroom/20050920a.html

12. UA HiRISE camera links fresh Mars gullies to carbon dioxide. University Communications (University of Arizona), 29 Oct 2010. http://uanews.org/node/33357

13. Odyssey studies changing weather and climate on Mars. NASA press release, 8 Dec 2003. http://mars.jpl.nasa.gov/odyssey/newsroom/pressreleases/20031208a.html

14. Ravilious K (2007) Mars melt hints at solar, not human, cause for warming, scientist says. National Geographic Society, 28 Feb 2007

15. Sigurdsson S (2005) Global warming on Mars? RealClimate (online), 5 Oct 2005. http://realclimate.org/index.php/archives/2005/10/global-warming-on-mars/

16. Marcus PS (2004) Prediction of climate change on Jupiter. Lett Nat 428:828–831

17. Jupiter's little red spot growing stronger. Sci Daily, 14 Oct 2006. http://www.sciencedaily.com/releases/2006/10/061013122425.htm

18. MIT researcher finds evidence of global warming on Neptune's largest moon. MIT News, 24 June 1998. http://web.mit.edu/newsoffice/1998/triton.html

19. Stern SA, Trafton LM, Galdstone GR (1988) Why is Pluto so bright? Implications of the albedo and lightcurve behavior of Pluto. Icarus 75(3):485–498

20. Pluto is undergoing global warming, researchers find. MIT News, 2 Oct 2002. http://web.mit.edu/newsoffice/2002/pluto.html

21. Marsh N, Svensmark H, Christiansen F (2005) Influences of solar activity cycle on Earth's climate. February, Publication of the Danish National Space Center

11. Ice Ages and Long-Term Cycles

Perhaps the most celebrated – and generally acknowledged – astronomical cycles are those that presumably have had no appreciable effect on the climate warming of the late twentieth century. However, they are thought by many to be the drivers of climate on a time scale of the ice ages. Known as Milankovic cycles, they could represent some of the greatest of all external influences on Earth and its climate. Although still controversial, the theory behind them has garnered the support of perhaps the majority in the mainstream scientific community.

Because they are closely aligned with everything under discussion in this book, a discussion of these cycles is also mandatory in this context. However, in no way should it be implied that they are responsible for recent global warming, even though some arguments could possibly be made for some unseen yet incidental, and presumably very slight, short-term connections to them in combination with other such factors. Unfortunately, many individuals have tried to make untenable cases, blaming Milankovic cycles for the recent warming in an effort to disprove existing AGW theories. This only further illustrates our need to understand all of the specific influences on climate in order to separate them. The context and function of the large astronomical cycles is just as important to grasp as the short-term influences, and how they have affected Earth's climate historically – and, more important than ever, how they will do so in the future.

A. Cooke, *Astronomy and the Climate Crisis,*
Astronomers' Universe, DOI 10.1007/978-1-4614-4608-8_11,
© Springer Science+Business Media New York 2012

The Milankovic Theory

Milutin Milankovic (1879–1958), by profession a mathematician, spent his greatest energies, however, on developing his visionary theory of climate. This was based on the blend of mathematical and astronomical theories. From early in life, he believed it was possible to codify what was responsible for the ice ages, and the warmer periods separating them. He also saw the possibility of developing a mathematical code for calculating the degree and effect of solar radiation as it passed through the atmosphere to the many different regions on Earth. In short, it seemed to him that climate had a predictable order. The best-known theory that he finally developed involved Earth's many orbital and axial variations, which became known as the cycles that bear his name. After a period of interest shown by the scientific community, the theory was eventually shunned until being rediscovered in the 1970s, its significance and potential finally understood and recognized.

It was clear that from early in Milankovic's career he had a passion that would remain with him for the rest of his life. He authored a number of astronomically related studies, beginning with the earliest form of his famous theory:

- 'Contribution to the mathematical theory of climate' (1912)
- 'The schedule of sun radiation on the earth's surface' (1913)
- 'About the issue of astronomical theory of ice ages' (1914)
- 'Researching the climate planet Mars' (1916)
- 'Mathematical Theory of Heat Phenomena Produced by Solar Radiation' (1920)

Thus, it was clear his attention was already focused on extraterrestrial horizons from the start. We can see how a grand theory of climate, in particular, was where it was all leading. However, it was not until 1927 that his momentous theory on the ice ages had developed into a fully fledged entity, as the introduction to *Handbuch der Klimatologie*, entitled, 'Mathematical Science of Climate and Astronomical Theory on the Variations of the Climate.' This soon developed into a full book in 1930, the iconic *Mathematical Climatology and Astronomical Theory of Climate Change*.

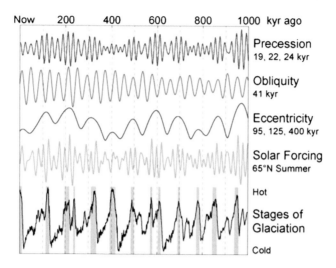

Fɪɢ. **11.1** Milankovic cycles (Graphic courtesy of Robert A. Rohde for the Global Warming Art Project)

Milankovic continued to refine and develop his theory for the next decade, with perhaps his most revered work being *'Canon of Isolation and the Ice Age Problem.'* A detailed study of planetary motions and the resulting forces between them, by the 1940s, however, his theories had already fallen into irrelevance, having to wait for a few more decades to become mainstays in modern science.

At the heart of Milankovic's theory are the analyses of planetary motions and tidal influences within the Solar System, all of which result in identifiable cycles on Earth in its climate over the ages. Milankovic followed in the footsteps of James Croll in the nineteenth century, who was first to theorize about orbital variations, and even the 100,000-year cycle of the ice ages. Figure 11.1 shows these identifiable cycles dating back over one million years, the resulting variations in solar forcing (from proxy records) as well as periods of glaciation. The approximate 100,000-year cycle is clear in the ice age record shown on this graph. However, the first three categories – precession, obliquity and orbital eccentricity – represent the primary astronomical ingredients of Milankovic's theory.

The keys to understanding Milankovic's theory are the variations in the positions and angles that Earth is placed in relation to

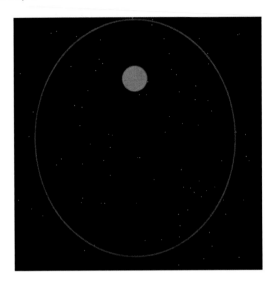

Fɪɢ. **11.2** Orbital eccentricity (scale exaggerated) (Graphic courtesy of NASA)

the Sun, and thus its degree of exposure to solar radiation. Because many variable factors work together, the formula for their interaction is complex. We should first consider the characteristics of Earth's orbit, which is far from a simple path:

- Small variations in its orbit from perfectly circular (slightly varying elliptically from between 0.005 to 0.058) influence the degree of warming Earth receives from the Sun. The Sun also is not to be found at the center of these elliptical orbits (see Fig. 11.2, elliptical orbit exaggerated), due to the gravitational influences of other planets, primarily Jupiter and Saturn. Although the amount of eccentricity might seem small, on the Earth-Sun scale it is apparently sufficient to produce a measurable difference. The reason for these orbital variations is also mostly due to the gravitational influences of other planets. The length of one full cycle of orbital eccentricity is approximately 100,000 years, although a more complex explanation is behind this cycle.
- Although a larger periodicity of 413,000 years may be seen to be stronger than the former (by its relative amplitude), in the simplest of terms, the combination with a number of other subcycles (the most significant being of 95,000 years and 125,000 years) has resulted in an *overall* observable cycle within that

larger one of approximately 100,000 years during the last million years (Fig. 11.1). It is this cycle that is seen to correspond to the rise and fall of successive ice ages.

- The orbital eccentricity results in Earth speeding up at perihelion and slowing down at aphelion, although it does not affect the length of the year. This is because orbital periods are fixed by the length of the semi-major axis, which corresponds to the radius of a circular orbit. However, these variations do result in slight adjustments in the lengths of the seasons. In the present orbital configuration, Earth receives almost 3.5% more radiation at perihelion than it does at aphelion, although at the maximum extremes of eccentricity the variation can be as much as 23%, quite a radical variation. This, all by itself, would be enough to trigger substantial changes in climate.

- For a relatively modest ellipse such Earth's present orbit subscribes, total radiation received (insolation) is not much different than that of a perfectly circular orbit, as well as being effectively averaged over the year. The same is not true for temperature over orbits of maximum eccentricity, however, as differences at both extremes may amount to several degrees Centigrade. At these times, this factor becomes highly significant in the resulting climate, especially when winters fall at the furthest reaches of the orbit.

Second, there are variations in the way Earth's tilt, relative to its axis (called obliquity), comes into play. Let us examine the obliquity of Earth's rotation (Fig. 11.3).

- This key variable, whereby the tilt of Earth's axis varies in motion, has been described by many as a "wobble." Almost everyone realizes that the angle of the poles is not 90° relative to the line subscribed by Earth as it orbits the Sun along its orbital plane. The period of variation of axial tilt for the full cycle is about 41,000 years, although that is also superimposed on a far larger cycle of about 1.25 million years. We can see that at one time the 41,000-year cycle was dominant but was superceded by what we now recognize as the 100,000-year cycle (Fig. 11.4). Similar to the 100,000-year cycle, the 41,000-year cycle incorporates other sub-cycles, in this case the most pronounced being the 19,000-year and 23,000-year cycles. Together, they result in what can be observed as the larger 41,000-year cycle.

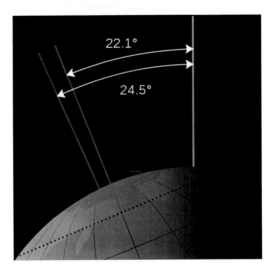

FIG. **11.3** Range of Earth's obliquity (Graphic courtesy of NASA)

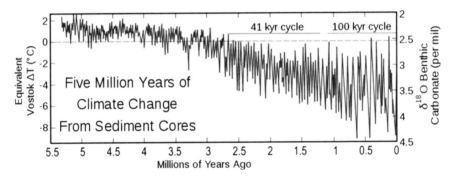

FIG. **11.4** The sudden change from 41,000-year to 100,000-year cycles (Graph courtesy of Dragons Flight and the Global Warming Art Project)

- Presently, Earth is at a mid-way position (at 23.5°) between the two extremes, and decreasing. Clearly both extremes will produce variations in radiation at the poles and equator, although globally, there is obviously the same amount of radiation reaching Earth. However, because Earth's largest landmasses lie in the Northern Hemisphere, the prospect of potentially greater or lesser accumulations of ice and snow become factors when this hemisphere is least favorably placed. This will again be the case about 9,000 years from now.

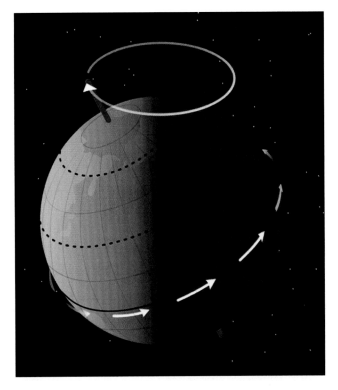

Fig. 11.5 Axial precession (Graphic courtesy of NASA)

A recent study (2009) by R. N. Drysdale et al. [1] concluded that obliquity could have significant consequences for climate. In this paper, a direct link was made to the culmination of the second last epoch (the Pleistocene), when about 141,000 years ago the change came to Earth's axial orientation. Presumably the author was implying a similar link to ice ages and other epochs.

Beyond obliquity, another motion, axial precession, plays a role (Fig. 11.5). Caused by the gravitational influences of the Sun and Moon, the poles slowly rotate around a near-circular path, which is also considered to affect the climate. During times when one axis is more favorably placed than the other with regard to the Sun, the effect will be one of greater seasonal extremes for one hemisphere than the other. Axial precession occurs in a cycle of approximately 26,000 years. However, although clearly capable of

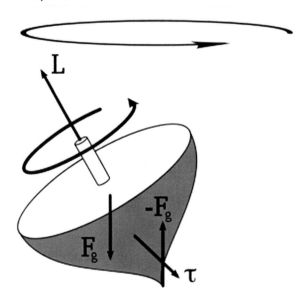

Fɪɢ. **11.6** Precession and obliquity (Graphic courtesy of Xavier Snelgrove)

having an effect on climate, we can see from the record that it is not a dominant cycle in the development of the ice ages. Regardless, this cycle seems most frequently associated with Milankovic theory.

We can compare this motion to the axial rotation of a spinning top or gyroscope. It is easy to separate obliquity from precession by visualizing this simple device in motion (Fig. 11.6). Although the axis rotates in a circle (precession), the angle at which it does so relative to vertical will determine the diameter that the circle subscribes (obliquity).

In addition to the seasonal variations that would be expected in a planet with axial tilt and an eccentric orbit (precession of the equinoxes), there is one other factor to consider (Fig. 11.7). The gradual accumulation of axial shift produces a change relative to the outside universe. Over a period of as little as 71.6 years, it will have shifted westwards around the ecliptic in the amount of 1°, completing the full circle in 26,000 years, (or 25,920 years, to be precise). Polaris will no longer be aptly named within a few thousand years. In about 10,000 years time, Vega will be the pole star!

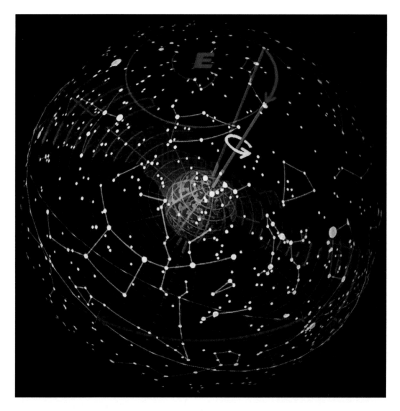

FIG. 11.7 Precession of the equinoxes (Graphic courtesy of Tau'olunga)

Strangely, some have commented that "no significant changes" are seen to be consequences of either precessional maximum extreme. This seems to be at odds with other sources.

In relation to Earth's axial orientation relative to its eccentric orbit, there is one other consideration. This results from a further complication known as apsidial precession. As was pointed out, during elliptical orbits, the Sun is not central to its orbit, an effect caused by the influences of Jupiter and Saturn. Over Earth's successive orbits, its perihelion and aphelion slowly rotate around the Sun, resulting in different hemispheres becoming more and less favorably faced to it, and appreciable variations in the precession of the equinoxes in the amount of about 4,000 years. This causes the 26,000-year precession to shorten to about 22,000 years and is

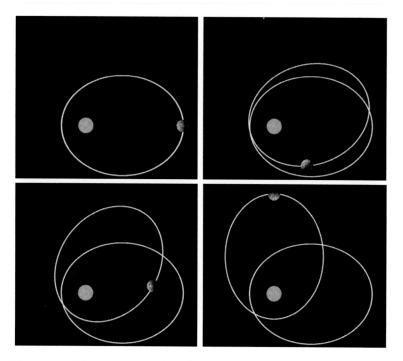

FIG. 11.8 Apsidial precession (Graphic courtesy of WillowW)

the reason we hear both numbers in reference to the precessional cycle – often without explanation.

Apsidial precession can be demonstrated graphically. At the present time, Earth is quite favorably placed, and thus winter (aphelion) in the Northern Hemisphere occurs most favorably in relation to the Sun's radiation. We can see in Fig. 11.8 how a few thousand years can alter this relationship quite appreciably as the ellipse slowly turns over upon itself (see Earth's orbit today relative to 5,000 years ago).

Remarkably, there is yet one other orbital variation that usually receives little comment. In regard to Earth's orbital plane, it also deviates from a 'flat' plane, up and down, over about 70,000 years. Not investigated by Milankovic, this has been the subject of some recent speculation. In relation to similar motions of the other planets, the effective motion also covers about 100,000 years, coincidentally, the same duration as the 100,000-year Milankovic cycle.

Ice Ages and CO$_2$

On the Skepticalscience website is the statement that CO$_2$ was not responsible for the warming from past ice ages but that it did cause amplification of the warming.

Such a definitive position is unfounded in any scientific realm. It is not possible to know from proxy records – all that is available to us from the ice ages – exactly what took place, other than to note increased levels of CO$_2$ that have historically accompanied warmer periods between the ice ages. Similarly, it not possible to cite decreases in CO$_2$ as being responsible for the ice ages, as some have done.

Muller and MacDonald [2] proposed that a meteoric collision might have produced an orbiting debris field that could have paralleled this period, especially since the sudden temperature shifts of the ice ages did not seem to correspond to smooth orbital motions. (The authors observed that the overly abrupt temperature peak in the cycle that does not seem in line with the more gradual effects that would be expected with normal eccentricity.) They suggested that a debris field might have produced the anomaly that appears in the historic record as Earth moves through it. Their findings also gave greater credence to variations of apsidial precession than to orbital eccentricity in the creation of the 100,000-year cycle. By comparing the motions of the 70,000-year cycle against the angular momentum of the Solar System (which approximates Jupiter's orbital plane), they found an excellent match to an alternative 100,000-year cycle that seemed to better fit the record. However, as it turns out, ice core records correspond closely to the timing of the Milankovic cycle and not to Muller's and MacDonald's hypothesis! Thus, Milankovic has prevailed.

The 41,000-Year Cycle Versus the 100,000-Year Cycle

Mainstream research has shown that the precessional variations, occurring within the 41,000-year obliquity cycle, should exert stronger influences on the climate than either the 41,000- or the 100,000-year orbital cycle. The fact that over the last million years the latter has showed up most strongly in ice core records (which revealed the ongoing cycles of ice ages of the same periodicity) has posed something of a problem for the theory. We can see that prior to this time, over about a million and a half years, the 41,000-year cycle is clearly dominant (see again Fig. 11.4). Only then, about a million years ago, does the 100,000-year cycle show up, which would be expected to be far less important as regards variations in incoming irradiation.

There have been several speculations that have been put forward as explanations. William Burroughs, in his excellent book, deals with this apparent contradiction quite thoroughly and convincingly; it seems reasonable. Briefly, it revolves around the theory developed by John and John Z. Imbrie of Brown University, Rhode Island [3], that utilized the geologic record of total land ice in preference to using the theoretical. Apparently climate models are roughly in agreement with the Imbries' finding that the 100,000-year cycle became dominant only after a critical mass of ice had been reached. A proposed 15,000-year cushion of delay allowed an accumulation of added ice to exceed the amount that melted between each ice age (despite the rapid retreat of ice after each period ended that is clearly evident in ice core records).

Continuing Controversy

However, as in all things associated with the climate, there have been continuing controversies regarding Milankovic's theories as well. It all began when isotope readings were sampled from a deep location in a Nevada aquifer (Devil's Hole). A 17,000-year discrepancy between the 141,000-year date given to the culmination of the second last ice age (as well as timings proposed by Milankovic's theory) and those of the samples was noted. This opened a new area of controversy, and the disputes continued with numerous competing papers, variously in defense of or against Milankovic's theory:

- In a 1992 article in *Nature*, Wallace S. Broecker concurred with the problems posed in relation to the specific historical timings predicted by Milankovic's theories [4].
- In direct response Cesare Emiliani [5] defended the theory, in claiming the reliability of such markers from the transitions between ice ages were questionable.
- Naturally, this was not the end of the discussion. In a follow-up, Landwehr et al. [6] accused Emiliani of biased research, whereby only data that was helpful to his case was allowed into his research. The authors stated that where events did not fit his criteria, he gave them less or no weight in his paper, effectively and selectively supporting only his case.

Regardless, the cause and effect remains in doubt, because others consider that the findings of 'Devil's Hole' may have been caused by entirely different historic factors. There has been continuing speculation about the Milankovic theory ever since, including the theories put forward by Muller and MacDonald as alternative explanations.

Whether time eventually proves Milankovic right or wrong, his theories still are embraced as the commonly ordained mechanism for the ice ages. Just as the conventional wisdom has, by and large, embraced AGW as the overriding cause of recent warming, Milankovic's theories have been similarly held as the correct ones at this time. Regardless, the continuing saga does further reveal the many counter-arguments that exist beneath the surface, just as we have seen with other theories proposed in practically every other aspect of climate research.

As may be seen by spending any time with the various articles and websites dealing with astronomical and solar cycles, the array and variety is almost endless. Other papers, such as that by Maya Elkibbi and José A. Rial [7], looked at some of the more familiar questions and tried to deduce an overriding formula to explain them. They found that:

- The 100,000-year cycle could be reproduced by a critical threshold of CO_2 being reached, despite a meager 1% increase in the resulting incoming solar radiation.
- That the 100,000-year may not be the blending of several other cycles as proposed, but orbital eccentricity may play a role after all by modifying precession.
- That the 100,000-year cycle may act as a non-linear amplifier of other frequencies, yet to be determined.
- That frequency modulation by the 41,000-year cycle of the larger 413,000-year cycle could possibly explain the 100,000-year cycle – just as a high-frequency 'carrier' modulates a lower frequency-modulating signal in FM radio.
- That increases in naturally occurring CO_2 rising to a specific level and later declining could have acted as the catalyst for each ice age, by allowing more polar ice to survive each cycle. This is somewhat in line with the Imbries' position [3].

The study concluded that any astronomical factors must be exceedingly complex combinations of effective forcings yet to be understood, in order to act together and be capable of making weaker cycles dominant. The authors found hard confirmation of Milankovic's theory scarce, but clues from astronomical sources plentiful. They concluded that the Milankovic theory is likely sound, but that it apparently varies and responds to all other Milankovic cycles in a non-linear (chaotic) pattern. Various alternative possibilities were raised, but suffice it to say, none of them has found universal acceptance, and Milankovic has still survived.

There are many other theories and analyses as well, and it will be up to the reader to try make sense of them if desired, such as a 1,800-year oceanic tidal cycle proposed by Keeling and Wharf [8]. Far from solar-oriented, the article proposed that lunar-Earth tidal forces of the gradual shifting in declination of maximum tidal forces over hundreds of years would produce a maximum effect every 1,800 years. The authors argued that we are presently in an upswing that would continue in an uneven manner for several hundreds of years. The most important part of this paper makes it clear that rapid climate change such as is presently under discussion may yet prove to have a hidden cause. Certainly it is unrelated to the longer-term scenarios of Milankovic.

The Law of Universal Gravitation

Isaac Newton's law of universal gravitation states that:
    Every object of mass in the universe attracts every other object of mass with a force that is directly proportional to the product of their masses and inversely proportional to the distance between them squared.

As one can see, prospect of known or yet unrecognized forces, even simple gravitational ones, playing more than an insignificant role in climate change are far from easy to dismiss. The variety of plausible-sounding theories is endless. Only time will tell which theories will survive the test of time, but even then, some researchers will surely question whether everything in play has been studied, or is yet even known. But the search continues by those still unconvinced that the correct formula has yet been found. However, the continued shredding of fine – even when relevant – minutia is far beyond the purposes of this book, and certainly beyond illustrating the general reasoning and precedent behind Milankovic's theories.

Perhaps a telling conclusion to this section might be the relatively early study (1976) by J. D. Hays et al. [9] that found precise and directly parallel evidence in ocean floor sediments for Milankovic's primary 41,000, 22,000 and 100,000-year cycles, along with firm conclusions that the theory held convincingly, despite the many challenges it has since received. For want of another explanation timed so well to these periods, it is still difficult to imagine another theory replacing it, although future refinements are still needed to explain some of those questions presently unresolved.

References

1. Drysdale RN, Hellstrom JC, Zanchetta G, Fallick AE, Sánchez Goñi MF, Couchoud I, McDonald J, Maas R, Lohmann G, Isola I (2009) Evidence for obliquity forcing of glacial termination II. Science 325:1527–1531
2. Muller RA, MacDonald GJ (1997) Origin of the 100 kyr glacial cycle: eccentricity or orbital inclination? Proc Natl Acad Sci USA 94: 8329–8334
3. Imbrie J, Imbrie JZ (1980) Modeling the climatic response to orbital variations. Science 207(4434):943–953
4. Broecker WS (1992) Upset for Milankovitch theory. Nature 359: 779–780
5. Emiliani C (1993) Milankovitch theory verified. Nature 364:583–584
6. Landwehr JM, Winograd IJ, Coplan TB (1994) No verification for Milankovitch. Nature 368(6472):594
7. Elkibbi M, Rial JA (2001) An outsider's review of the astronomical theory of the climate: is the eccentricity-driven insolation the main driver of the ice ages? Earth Sci Rev 56:161–177
8. Keeling CD, Wharf TP (2000) The 1,800-year oceanic tidal cycle: a possible cause of rapid climate change. Proc Natl Acad Sci USA 97(8):3814–3819
9. Hays JD, Imbrie J, Shackleton NJ (1976) Variation in the Earth's orbit: pacemaker of the Ice ages. Science 194:1121–1132

12. Cosmic Crisis

Milankovic's theories were not the only ones to emerge out of the search for hidden clues to the large climate variations and epochs that have taken place throughout Earth's history. However, it concerns only the domain of the Sun. Two related theories, with implications from far more distant places than any within the Sun's realm, also have been at the heart of no less bitter a controversy and dispute than any we have encountered before. Probably the most esoteric and contested of them all, they link us to the very heart of the galaxy, involving the effects of galactic cosmic rays.

Although the proponents presented detailed cases and analyses, these theories have had a rough ride, not unlike the furor surrounding the ACRIM findings but on a grander scale. Although some debunked the hypotheses, not all researchers have reached an opinion, maintaining that no final resolution has been determined to their satisfaction. Meanwhile, the proponents have vigorously defended their turf.

The possible link of cloud formation to galactic cosmic rays has been at the center of intense speculation on and off for many years. In 1959 Edward Ney, who had been studying them since the end of World War II, published his landmark study, 'Cosmic Radiation and the Weather,' in which the connection apparently was first made a high level in the scientific community [1]. Indeed, Burroughs referenced other pioneering work in cosmic radiation being undertaken by Ralph Markson of MIT; this has continued up to the present. Markson's primary research, though concerned with the possible connection of cosmic rays to thunderstorms, has developed further into examining possible electrical connections to global warming itself [2].

A. Cooke, *Astronomy and the Climate Crisis*,
Astronomers' Universe, DOI 10.1007/978-1-4614-4608-8_12,
© Springer Science+Business Media New York 2012

251

The Galactic Findings of Henrik Svensmark

In 1997 came a theory that would jump-start all the research and controversy that was to follow. Henrik Svensmark and Eigil Friis-Christensen's 'Variation of cosmic ray flux and global cloud coverage – a missing link in solar-climate relationships' (referenced in Chap. 4) proposed that cloud cover was triggered by cosmic ray flux (CRF) [3]. They based their assessment on data from the International Satellite Cloud Climatology Project (ISCCP), the Defense Satellite Meteorological Program (DSMP), and the Nimbus-7 program. Although it is fairly well established that the Sun's magnetic field has profoundly deflective properties against incoming cosmic rays, Svensmark and Friis-Christensen claimed corresponding global temperature variations were better synchronized with CRF variations than direct solar activity. Thus, a link to climate and the solar magnetic field might prove to be the elusive link to climate change – the possible indirect warming mechanism many climate scientists had been looking for.

Cloud Formation and Cosmic Rays

Cloud formation cannot occur without a seeding mechanism. In the case of the atmosphere the seeds are fine particulates in the form of aerosols that act as the trigger to begin the process of the condensation of water vapor into droplets. With better cloud monitoring and precipitation records over solar cycles, climate models would benefit greatly. Bear in mind that cloud formation in the lower atmosphere is seen as quite different from that of the stratosphere, where ice clouds form by different processes and could not thus be tied to cosmic rays.

The theoretical process of cosmic ray interactions with the lower atmosphere to form aerosols can be summed up as follows:

1. Photochemical reactions change atmospheric sulfur dioxide to sulfuric acid.
2. Cosmic rays ionize sulfuric acid into ultra-fine ionized gas particles.

3. These particles gather and clump together.
4. The clumps condense further into larger cloud condensation nuclei.
5. Water vapor is attracted to these larger highly charged aerosols and condense into droplets.
6. Clouds begin to form.

Svensmark and Friis-Christensen tied the 11-year solar cycle to what they claimed was an observed 3–4% century-long variation in cloud cover caused by changes to the strength of the solar magnetic field. Recognizing that clouds of all types resulted in an overall net cooling effect of the lower atmosphere (by reflection) than heating (by greenhouse warming), they theorized that incoming cosmic rays led to ionization of gas particles in the atmosphere. These could serve as aerosol catalysts for low altitude cloud formation. Thus, during times of the Sun's greatest activity, the increased solar magnetic field would reduce the amount of cosmic rays able to penetrate Earth's atmosphere and lead to less cloud formation and higher temperatures.

Cloud Categories and Relative Altitudes

Cirrus *(highest)*

Cirrostratus

Cirrocumulus

Altostratus

Altocumulus

Cumulonimbus *(reaching from ground level even above cirrus clouds)*

Nimbostratus *(reaching from ground level into the upper atmosphere)*

Cumulus

Stratocumulus

Stratus *(low clouds and fog)*

Over the course of the twentieth century an ever-stronger solar magnetic field (through growing solar activity) had increased the Sun's cosmic ray shielding properties; thus, less CRF-generated cloud had led to a reduction in the total cloud component of Earth's composite albedo. The evidence seemed to fit the estimated doubling of the solar magnetic field to the estimated 3.5% decrease in cosmic ray flux, with a parallel increase in global temperatures. However, the new theory was soon to spawn the latest great storm of controversy in the climate wars, as questions and challenges immediately arose from what was loosely termed a problem with 'the satellite viewing angle' of the three sets of data utilized from Nimbus-7/ISCCP/DMSP, Nimbus-7, interestingly being the same satellite that was tied to the ACRIM controversy.

In 1999, Jorgensen and Hansen (not James Hansen!) for the IPCC TAR challenged the veracity of the findings, claiming that none of the necessary proof existed to support the theory [4]. Svensmark and Friis-Christensen responded that just because they could not conclusively state that their CRF theory was beyond doubt did it necessarily rule it out [5].

Regardless of the early negative response, Svensmark followed up on this paper over the next few years with a flurry of new research papers along similar lines (sometimes with his colleague, Nigel Marsh) [6]. These included references to Milankovic's orbital theories, ENSO cycles, and past solar activity cycles, amongst other factors. Amongst them, perhaps the case for cosmic ray influences was stated most effectively with, 'Low Cloud Properties Influenced by Cosmic Rays', in which Marsh and Svensmark focused on the link of CRF to *low* altitude cloud formation – the densest and most reflective of cloud layers.

Having then reconstructed the historical low cloud cover (LAC) over the century, Marsh and Svensmark deduced an increased warming from the reduced cloud cover to be about 1.4 W/m^2, an amount approximately equal to that which had been theorized due to anthropogenic causes. They continued further, going back in time over thousands of years, concluding there was a reasonably good correlation with climate and cosmic rays, as deduced through carbon 14 records. It should be pointed out again that the authors did not state their case as a certainty but rather as a foundation for further research. They also did not rule out anthropogenic contributions. This is an important distinction – quite a common position among scientists who have explored alternative theories and who are often unfairly accused of denying any warming effect from anthropogenic causes.

However Svensmark, by this line of research, and also presumably through his spirited public defense of his controversial research findings, had become a 'darling' of some, and thus, by default, increasingly likely to come under heavy fire, although he did receive some early support from a number of other researchers [7]. Convincing to Lockwood et al., for example, was how the observed temperatures seemed to fit the relatively well-established solar magnetic flux for the period under review (1964–1995) [8].

The Other Side of the Coin

A remarkably damning paper by Peter Laut found very little to like in Marsh's and Svensmark's work [9]. Laut's analysis led him to conclude that the data had been misinterpreted, and that some of the newly sensational findings of Marsh and Svensmark were at odds with the facts.

Laut first took issue with the switch of the description of 'total cloud cover' from Svensmark and Christensen's findings to 'low cloud cover' in Marsh and Svensen's subsequent work. Since low clouds are dense and highly reflective, and thus have very different implications for the influence of cosmic rays, it was an important distinction. Further, Laut could find little correlation between the two sets of satellite data in many instances, and created a 'corrected' graph by removing the DMSP satellite data (that he claimed did not represent total cloud cover) and adding new ISCCP satellite data. Here cosmic rays and total cloud cover seemed to be widely diverging after 1992.

In citing other studies that found no correlation whatsoever between cosmic rays and cloud formation, additionally, he found issues with the timing of cosmic ray flux (CRF) and cloud formation, which he stated should have been virtually instantaneous [10]. Commenting on the difficulties of differentiating between types of cloud cover from the satellite data, Laut continued with critiques of a number of other similar papers, especially the work by Lassen and Friis-Christensen, again faulting the interpretation of data [11].

Fraudulent Climate Science?

To form any kind of opinion, we need to look at contrary views, because the data behind such esoteric concepts is liable to have numerous interpretations. Svensmark had done his best to explain the limitations of the data at the time of his research some 3 years earlier [12]. Now he considered he had come under personal attack, effectively having been accused of committing scientific fraud. However, he went on to justify why and how it was perfectly correct to include it, and how the change from 'total cloud cover' to 'low cloud cover' came about with the emergence of new information. He considered his paper had put an emphasis on the need for further research, since an ideal monitoring system did not exist at the time of writing. He also insisted that he had taken into account and expressed all possible reservations that could be drawn from the findings, providing full transparency of the methodology utilized for all subsequent researchers.

His protests were in vain, however. Laut would attack what he saw as an inappropriate response being published on the Danish Space Research Institute's website – where Svensmark worked – and not in a peer-reviewed journal in which it would be taken more seriously within the scientific community [13]. Laut suggested that this route might have been chosen because it might not stand up to peer review. Svensmark may have felt that he had no choice but to act in the quickest possible way, given the stakes for his future credibility in the wider community. Regardless, having made his point, Laut then proceeded to commit the same 'offence,' publishing this second piece on the website of the Technical University of Denmark, his own workplace!

Laut further accused Svensmark of manipulating the text of his (Laut's) paper. However, in light of what appeared to be a related accusation by Svensmark about Laut's analysis, Laut also suddenly seemed to be on the defensive, justifying why he was not obligated to check all of Svensmark's results. In highlighted wording for added emphasis, Laut once again made it clear that he considered that Svensmark had misunderstood the very meaning of the DMSP satellite data. He returned to the distinctions in cloud cover between 'low' and 'total' again and again, ramming home his unchanged view that the data had been misused, misapplied, and misunderstood.

Stalemate

All the materials warrant careful study if one wants to fully grasp what was at the center of all the fuss. One's interpretations and conclusions will depend on many things, some of which may be mired in one's own mindset. However, there can be no doubt that this was an example of climate change acrimony at its most ugly and personal. One wonders how these researchers would handle a situation should they find themselves in the same room at some social function! The argument was thus left in limbo, with some scientists still intrigued at the possibilities. However, it was frequently stated in many blog comments that the theory had been debunked and discredited, so for the time being, at least, the theory had taken on enough water to have become somewhat bogged down under its own weight.

Regardless, other articles emerged from time to time that kept the door open, such as one by E. Pallé, who ventured to dip his toe into the icy water, commencing with the prospect that satellites may have "artificially induced" the apparent extent of low cloud cover [14]. His concluding remarks left open the distinct possibility of cosmic ray influence on low cloud formation, and stated with a fair degree of confidence that Sun/cloud relationships may have played a large role in pre-industrial climate variation – not an insignificant position to take. Svensmark's work continued concluding there was 'conspicuous' evidence of the effects of variations in solar activity in the creation of clouds and aerosols, as evidenced by Forbush Decrease events (FD's) – the immediately measurable decreases of cosmic rays striking Earth [15].

Low clouds are found from near the surface up to 6,500 ft (2,000 m). Typically this also includes cumuliform and stratiform category types. When low stratiform clouds contact the ground, they become one of many categories of fog. These include radiation and advection and surface-generated types that do not form from such stratus layers.

Meanwhile, another bombshell had already dropped and would extend the argument and controversy further and into another realm (if not quite another galaxy) entirely....

Nir Shaviv and the Galactic Arm Connection

Sure to create additional dispute was the daring research of Nir J. Shaviv, one of the new generation of astrophysicist/climate researchers. Though born in New York, he is currently based at Hebrew University of Jerusalem. Shaviv is well known for his negative assessment of CO_2-based warming, the widespread acceptance of which he, too, blames on the media. Like Svensmark, Shaviv, too, would become a target. However, Shaviv had also proven to be a formidable adversary, having built up considerable respect from worldwide colleagues with seemingly impeccable credentials and scholarly brilliance.

He was soon to become doubly famous and controversial for his 2003 findings with fellow researcher, Ján Veizer, for placing an upper limit for temperature increases on a future doubling of CO_2 atmospheric concentrations [16]. Here, they put it at about one third that estimated by the IPCC 2001 climate model: about 0.5–1.9°C, versus about 1.5°C to about 5.5°C. Thus Shaviv's and Veizer's average produced a rough estimate of about 0.75°C. Among his explanations for the upward temperature trend since then, he considered that the oceans played a significant role in tempering short-term effects of the Sun, which has delayed the reaction to a cooling Sun (Long Term Persistence–LTP, again).

Shaviv has long considered the *Medieval Climate Optimum* and the *Little Ice Age* were certainties, linking them to $_{14}C$ flux via cosmic rays being deflected by the varying solar magnetic field, or possibly the photochemical reactions of UV in the upper atmosphere. Shaviv already had made no bones about the Sun-climate connection, further correlating the increasing temperature curve to solar variation rather than the monotonic anthropogenic increases [17]. However, with the view that the 0.1% variation is insufficient in itself to directly affect the climate significantly, he thus indicted theories of warming process by that means.

Forbush events – the immediately measurable decreases (by means of particle detectors) of cosmic rays striking Earth – become significant in Shaviv's thinking, following a coronal mass ejection from the Sun, accompanied by sudden bursts in the strength of the Sun's magnetic field. In Shaviv's theories, we find frequent mention of Forbush events (especially in regard to the research of Tinsley and Deen [18]). Shaviv noted direct consequences to the climate, as well as variations in the CRF over the course of each 11-year solar cycle. Thus, he further speculated on the plausibility of CRF on cloud cover, in essence supporting the aerosol-CRF contentions of Svensmark. Shaviv must have known that he was reopening unhealed wounds, but obviously he felt this was an important part of climate research that had been relegated to the attic.

Proceeding further, he looked for possible cosmic links to twentieth-century warming trends. Specifically mentioning the role of greenhouse gases in Earth's recent temperature changes, he looked at periods during the century when *declines* in temperature were recorded. This was because he had logically deduced that increased anthropogenic gases could not have contributed to periods of cooling. Thus, with the period from the 1950s until the 1970s, as well as support from various reference sources, Shaviv believed he had correctly attributed a 1.5% increase in CRF to the 0.2°C cooling trend that took place at that time. He actually figured that cosmic rays would have had a greater effect on temperature than this, since the data included ocean temperatures, the expanses of oceans having greater heat storage capabilities than land.

The Galactic Climate Connection

However, the real crux of the paper came with the new theory that directly tied long-term periods of CRF, not merely to such local phenomena as just described, but to the passage of the Solar System through the spiral arms of the galaxy! This was to be at the core of a new theory that proposed that the great ice age epochs had a distant celestial origin and was the focus of a follow-up paper later that same year [19]. Shaviv argued, with the aid of carefully constructed diagrams, that CRF coincided with the geologically reconstructed periods of those ages, dating back 500 million years (the Phanerozoic period). Certainly these diagrams made things *look* clear enough. However, resistance to these ideas sprung forth very quickly and forcefully, especially (but not unexpectedly) from the AGW proponents.

We should also be sure to distinguish between the variations of CRF on Earth through solar activity, versus those that might come about through its placement in the galaxy. Many have theorized that cosmic rays are byproducts of supernova explosions, the consequences of which reverberate throughout the galaxy. Because stars that ultimately end their days in this spectacular fashion are inherently unstable, due to their core mass growing to exceed the theoretical limits (the 'Chandrasekhar Limit'), they will not have time during their relatively short existence to progress far in their orbits around the galactic core. Thus, many will meet their untimely demise not far from their place of creation – in the spiral arms, because obviously more supernova events will happen in regions where there are greater stellar concentrations.

Supernovae and the Chandrasekhar Limit

This is the maximum mass that a white dwarf star may maintain or attain ('dwarf' being a terminology that refers to all main sequence stars before they attain red giant status) before becoming unstable. It was named after Subrahmanyan Chandrasekhar, who formulated this stellar constant in 1930. The numerical value of the limit is about generally accepted to be approximately 1.4 solar masses.

Also, galactic structure is not all that it might seem. The arms are more akin to compression zones than the familiar firework pinwheel arms, where stars periodically become more densely packed together in waves during the course of circumnavigating the galactic center. Not everyone is aware that spinning motions of the galactic core do not form the galactic arms; pinwheels might look similar but actually have nothing in common. In fact, galactic arms are the product of density waves. There is a considerable density of stars in between the spiral arms as well, but being far more tenuous populations they are much less obvious. However, any experienced observer knows well the luminous galactic 'halo' that immerses the entire galactic structure. Halos are comprised of stars, the luminous product of which we can readily see in eyepiece views in the brighter examples. CCD imagery makes this even clearer, of course.

Shaviv theorized that CRF could be directly traced back through the historic record to an approximate historical synchronization of Milky Way's arm placement with the epochs. His theory was reliant on known models of its structure, at that time believed to be a very large, four-armed barred galaxy. When the Sun passes through one of the arms, an increase in CRF would be anticipated. In pursuing his theory, remarkably, he found uncanny matching of the calculated passage of the Sun through these spiral arms and the onsets and declines of these past epochs. He theorized that increased CRF during these crossings resulted in increased low cloud cover and thus the onset of declining temperatures.

Interestingly, Shaviv referenced other earlier studies that had theorized galactic connections, including one by Hoyle and Wick-ramasinghe that theorized climate warming should occur if the Solar System happened to pass through a cloud of interstellar matter, due to increased solar irradiance [20]. Shaviv listed several other possibilities, too, some of them decidedly serious for life on Earth. All of this is the stuff of further research, and even were any of it occurring now, this could not be the cause of present-day climate change because it would take place over far greater lengths of time. This distinction is important to know, since so many have tried inaccurately to tie the recent twentieth-century warming to such esoteric concepts, when, in fact, they are surely unrelated.

Naturally, those opposed to conventional theories rejoiced, even if such findings could not be tied to the recent short upward spike (at least, in these galactic terms) and present temperatures on Earth. Many chose to ignore this important distinction. It was a complex presentation and analysis indeed, thoroughly reasoned and appearing to produce compelling evidence to support it. With a calculated timing for galactic arm crossings was 10^8 years, Shaviv thus believed that he had been able to show which arms were crossed at what time in history, and the parallel ice ages that resulted from them. The implications of this amazing and innovative piece of cosmic thinking were huge.

Exploring the CRF link as far as he could, Shaviv then estimated that the larger signal was of the order of approximately 143 million (±10 million) years, which corresponded reasonably closely to the periodicity of the spiral arms, estimated to be around 134 million (±25 million) years. He had also made mention that different zones within the arms, such as stellar nurseries, could be expected to produce significant "intrinsic" increases in CRF, along with periods of short term cooling. However, his theory was beautifully conceived for a four-armed barred spiral. Somehow it was all too perfect.

The Wrong Galactic Model!

It was only a matter of time before a new model of the spiral formation of the Milky Way was theorized, especially since the four-arm concept was not proven, and was only considered the most likely form of the galaxy according to knowledge of the time. Even Shaviv had acknowledged that. In 2008, a revised projection of the Milky Way appeared from NASA, whereby it now showed only *two* main arms. This certainly was problematic to the CRF spiral arm theory and was immediately seized upon by Shaviv's detractors to discredit it. However, the structure of the galaxy as revealed now is not altogether a simple two-armed affair, since there are other lesser arm components – the spurs. Spurs are more transitional than the arms, although even the great spiral arms themselves only have a life span of 900,000–1,500,000 years. Thus, the galaxy is an ongoing work in progress, and appears to be more complex from the 'overhead' view than many others.

Regardless, the new two-armed theory must certainly have been a setback for Shaviv, since many claimed that the findings effectively eliminated his theory. However, it is perhaps a little too soon to rule out all possible connections to the spiral arms; we still do not know the Milky Way's precise form with certainty, and only time will determine it. Only then will it be possible to see if any possible waves of CRF could be related to the great epochs. Is it unreasonable to surmise that perhaps the 'spurs' might themselves be sufficiently dense to initiate some degree of increased,

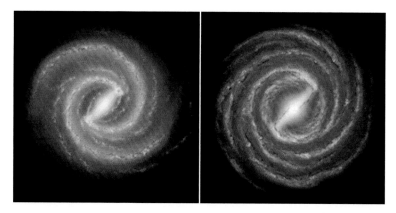

FIG. 12.1 The two Milky Ways: the 'New' and the 'Old' (Graphic courtesy of NASA/JPL-Caltech)

moderately long-term cosmic ray incidence? Clearly, Shaviv has seen no reason to yield any ground (Fig. 12.1).

Regardless of how the Milky Way's form ultimately turns out to be, at present there are still four named arm/spur components to the galaxy, as shown in Fig. 12.2, which also shows Earth's position and orientation within it. Could these be sufficient to support Shaviv's theories?

Although the furor swirling around this latest expanded view of Svensmark and Friis-Christensen's original CRF theory would continue in short order, it should be noted that Shaviv went to considerable pains to acknowledge that much within his theory was just that – theory – and yet to be proven. In referencing various caveats, it was not as if he had stated some kind of heretical position, or had not recognized that its circumstances were far from settled, much less even known. He had also acknowledged that the spiral arms were not all created equal – well before the newly configured Milky Way map had appeared, and had even referenced its 'spurs' (relatively short-lived structures) as opposed to the spiral arms. Undaunted, Shaviv would follow his original paper with another with co-author Ján Veizer, referenced earlier, continuing the spiral arm theory [16].

Not all of the critiques demonstrated comparable quality of analysis to Shaviv and Veizer's work [21]. Regardless, some members of the scientific community pointed out that Shaviv's original study had based its findings on 50 iron meteorites coordinated into the theory of galactic spiral arm crossings. However,

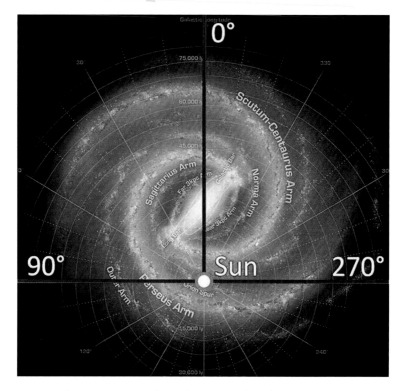

FIG. 12.2 Earth's position in the galaxy, comfortably within the Orion Spur lying between the Sagittarius Arm and the Perseus Arm (Graphic courtesy of NASA/JPL-Caltech, and Brews O'Hare (grid))

reliable dating of such meteorites, and the interpretation of the 143 million (±10 million) year periodicity from a variety of only approximate and mixed results, were not agreed by all, nor the deduced ages demonstrating anything near enough the type of periodicity that would be expected in galactic arm crossings. Some conceded that CRF might have some place in climate epochs of the ages, but that CO_2 remained the main driver. Further proxy evidence from over the prior 15 years was utilized without success in corroborating any principal connection of CRF to climate.

Other reservations from Shaviv's findings included:

- The premise that as much as 66% of recent warming was not attributable to AGW.
- The reconstructed sea surface temperatures (SST) over hundred of millions of years.

- His 'arbitrary' change in time scaling to coordinate with CRF at low altitudes.
- The projection of warming that would result from a doubling of CO_2, calculated at an average of only about 0.75°C.
- The reference to 'an incorrect' quote of IPCC projections for the worst-case scenario at 5.5°C, instead of 4.5°C.

Meanwhile, Shaviv and Veizer, claiming that the very uncertainties *they themselves had raised* had been used against them, provided:

- The methodology of dating the meteorites, along with the dependability of their findings, supporting the projected CRF variability, and their temperature findings of past epochal radiative forcings.
- References to the table in which uncertainties in the calculating timings of the crossings of spiral arms were addressed, including reservations about the Milky Way's 'Norma' Arm.
- Influences of orbital forcing (Milankovic cycles).
- 'Tuning' methods for ill-synchronized sedimentary deposits to make the data sets correspond.
- Justification for their superior margin of error of the delayed response to galactic arm crossing.
- Clear CRF parallels.
- Justification for the discrepancies between the temperature range referenced for CO_2 increases in IPCC climate models and the stated projections used in the TAR (for which *they* had been criticized) [22].
- Further detailing and graphics [23].

Looking for Middle Ground

As the debate continued, Shaviv and Veizer appeared to be looking for common ground even though they clung to their original conclusions [24]. Allowing that the 'pH effect' discussed at length (by Royer et al.) might have distorted the data in reconstructing past temperatures as being too high, it was acknowledged that good research may indeed have improved the data [21]. Shaviv continued to detail his theorized link between CRF, low altitude cloud cover and climate change, especially with regard to the epochs. In this analysis, he included as much from the Royer et al. study as seemed relevant. His conclusions were that anthropogenic warming was responsible for no more than 0.14–0.36°C of the total warming since 1900, with theorized sources and actual temperature observations being remarkably compatible. However, the key to retaining the intact spiral arm crossing theory depended on Marsh's and Svensmark's fundamental theory.

Similarly objective was the work of several other researchers, with the findings on CRF and the spiral arm crossings often inconclusive:

- The link between cosmic rays induced ionization (CRII) and low altitude cloud cover (LAC) was often upheld, not only concurrent with the 11-year solar cycle but also being dependent on latitude.
- Towards the poles, the effects of CRII and increased low cloud cover seemed most pronounced, exactly the opposite of what one would have expected, had solar irradiance been the primary driver.
- Further conclusions indicated that the influence of CRII at mid-latitudes far outweighed those related to solar activity, and that key information was being lost due to the averaging of all regions together.
- Some regret was also voiced that the influence of cosmic rays was largely absent in climate models [25].

Thus, Shaviv and Svensmark found that they had at least some common ground with other researchers, in addition to what existed between them. Along with F. Christiansen, Svensmark further would conduct new research on the cosmic ray, solar wind and UV links and would even feature Shaviv's spiral arm theory prominently [26].

Death Knell for a Theory?

As the first decade of the twenty-first century was drawing to a close, some newer studies with updated methodologies would pour cold water all over the CRF theory [27].

Among the more negative outcomes of the new studies:

- Recent warming was not found tied to CRF, although an undecided position on long-term hypotheses did little to keep Shaviv's spiral arms concept alive.
- One of the key personalities originally to concur with the CRF theory apparently had changed his mind. In this particular review, conclusions from this and other research appeared to discount the theory on every level.

However, at least some highly detailed and comprehensive studies were being undertaken, exactly the kind that Svensmark and Shaviv had suggested was needed to explore the possibilities further. However, for those who had invested so much time and effort, to say nothing of their reputations, it seemed that the newer research had trumped the older. Maybe the time had come to throw in the towel.

The CLOUD Experiment

CERN is the acronym for the European Organization for Nuclear Research, responsible for, among other things, the Large Hadron Collider [28]. In 2009 the organization initiated research into the core of Svensmark's argument – to see if cosmic rays could produce the results he predicted and had so energetically championed. The program was named the CLOUD experiment (Cosmics Leaving OUtdoor Droplets) and involved the use of a high-energy physic accelerator (the Proton Synchrotron) and a cloud chamber [29]. The chamber simulated the atmospheric environment, and the accelerator was used to bombard it with artificial cosmic rays. It was hoped that it might settle the argument once and for all, although for many researchers, the answer was already a foregone conclusion.

On August 29, 2011, 1 month to the day before the completion of the first draft of this book, CERN announced that the first results of 2 years of work had been published in *Nature* [30]. They were startling by any standards:

- First it was determined that the sulfur-based vapors (mostly sulfuric acid H_2SO_4), always thought responsible for most lower atmosphere aerosol production, actually accounted for very little of it.

- Second, clear linkage of cosmic rays to aerosol production was established.
- Third, Svensmark's theory of clusters of sulfur compound particulates gathering into aerosols large enough for cloud formation being dramatically enhanced by CRF was borne out, by a factor of *at least* ten times.

However, a little caution was raised. It was apparent that ammonia-type compounds are also necessary for low altitude cloud formation. CLOUD had showed that even with cosmic ray bombardment of sulfuric acid, ammonia and water vapor alone were still insufficient in themselves to account for aerosol production – in the absence of some other unknown compounds. Additional supporting information is available from the CERN website [31]. Next on the agenda for the CLOUD experiment is therefore to find such a compound(s), should this indeed be present.

Could it be that Svensmark, perhaps the most ridiculed of the recent alternate thinkers, suddenly had been vindicated, along with more than a few other outspoken scientists who backed him? Perhaps even the spiral arm connection will prove to have some value as well, and Shaviv could yet have his day. Only time will tell, of course, where the continuing research will bring us, and what it all means for climate science, especially regarding all similarly related research and hypotheses.

In Conclusion

Certainly some of the theories that we have examined in this book are controversial and far from universally accepted or even respected, but many of their advocates are leading figures in research, and their studies are fascinating at the very least. Regrettably, the information is not well promulgated at this stage. The lack of general recognition within the larger climate community may speak more to unfamiliarity than lack of validity, or any desire to suppress it, although even this may be true in some cases.

Cosmic Rays and Climate

On its website, the Environmental Defense Fund states that there has been no trend in incoming cosmic rays detected over the last 30 years, and that these rays have had "little" impact on recent warming.

These so-called facts are stated in a vacuum. And the term "little" is interesting, especially in lieu of actual quantifiable measurements. Are we supposed to discount cosmic rays, but then wonder what their actual effect might have been? No serious scientist would ever make such missteps.

Predictably, the ink barely had time to dry on the CERN press release (concerning the results of the CLOUD experiments) than *Skepticalscience* acted at light speed to dismiss the results, while not seeming to represent what actually had been claimed. Apparently this organization sees little purpose in further investigation and has already concluded that CERN has no case.

Whether this is, or is not so, as always, however, Canute will have the final word.

Should critics choose to characterize much of what is contained in these new studies as unproven, that is fine, except there are many positions and conclusions they may accept that are similarly unproven! Perhaps we will not have to wait too many years to finally come to a mutual understanding of everything. However, at least for now we should no longer be surprised at unusual findings that seem to fly in the face of the accepted norms of climate science. Perhaps we really don't know *all* that there is to know, and maybe the debate is really not over after all. Time and Canute will tell (Fig. 12.3).

F<small>IG.</small> **12.3** Earth observatory – top of the atmosphere (Image courtesy of NASA)

References

1. Ney EP (1959) Cosmic radiation and the weather. Nature 183:451–452
2. Markson R (2000) The global circuit intensity; its measurement and variation over the last 50 years. Bull Am Meteorol Soc 88:223–241
3. Svensmark H, Friis-Christensen E (1997) Variation of cosmic ray flux and global cloud coverage – a missing link in solar-climate relationships. J Atmos Sol Terres Phys 59(11):1225–1232
4. Jørgensen TS, Hansen AW (2000) Comment on "Variation of cosmic ray flux and cloud coverage – a missing link in solar-climate relationships". J Atmos Phys 62:73–77
5. Svensmark H, Friis-Christensen E (1999) Reply to comments on "Variation of cosmic ray flux and global cloud coverage – a missing link in solar-climate relationships". J Atmos Terres Phys 62:73–77
6. Svensmark H (1998) Influence of cosmic rays on Earth's climate. Phys Rev Lett 81:5027–5030; Svensmark H (2000) Cosmic rays and Earth's climate. Space Sci Rev 93:175–185; Marsh N, Svensmark H (2000) Cosmic rays, clouds, and climate. Space Sci Rev 94:215–230; Marsh ND, Svensmark H (2000) Low cloud properties influenced by cosmic rays. Phys Rev Lett 85:5004–5007

7. Landscheidt T (2000) Solar wind near Earth: indicator of variations in global temperature. In: Vázquez M, Schmieder B (eds) The solar cycle and terrestrial climate. Proceedings of the solar and space weather euroconference, vol 463. ESA SP, pp 497–500; Shumilov OI, Raspopov OM, Kasatkina EA, Turjansky VA, Durgachev VA, Prokorov NS (2000) Atmospheric aerosols created by varying cosmic ray activity – one of the key factors on non-direct solar forcing of climate. In: The Solar cycle and terrestrial climate. Proceedings of the solar and space weather euroconference, vol 463, ESA SP, pp 543–546; Tinsley BA, Yu F (2002) Atmospheric ionization and clouds as links between solar activity and climate. J Am Geophys Union 141:321–340

8. Lockwood M, Stamper R, Wild MN (1999) A doubling of the Sun's coronal magnetic field during the last 100 years. Nature 399: 437–439

9. Laut P (2003) Solar activity and terrestrial climate: an analysis of some purported correlations. J Atmos Sol Terres Phys 65:801–812

10. Kernthaler S, Toumi R, Haigh J (1999) Some doubts concerning a link between cosmic rays and global cloudiness. Geophys Res Lett 26:863–865; Wagner G, Livingston DM, Masarik J, Muschler R, Beer J (2001) Some results relevant to the discussion of a possible link between cosmic rays and Earth's climate. J Geophys Res 106: 3381–3388

11. Lassen K, Friis-Christensen E (2000) Reply to the article "Solar cycle lengths and climate: a reference revisited". J Geophys Res Space 105:27493–27495

12. Svensmark H (2003) Comments on Peter Laut's paper, "Solar activity and terrestrial climate: an analysis of some purported correlations". Danish Space Research Institute. http://docs.google.com.viewer?a=v &q=cache:EQUGKMOF7Lwj:www.space.dtu.dk/upload/institutter/ space/torskning/05_afdelinger/sun-climate/full_text_publication

13. Laut P (2003) Comments by Peter Laut on: Henrik Svensmark's "Comments on Peter Laut's paper, 'Solar activity and terrestrial climate: an analysis of some purported correlations'". J Atmos Sol Terres Phys 65:801–802; Svensmark H, Laut P (2003) Technical University of Denmark, 2003

14. Pallé E (2005) Possible satellite perspective effects on the reported correlations between solar activity and clouds. Geophys Res Lett 32:L03802

15. Svensmark H (2007) 'Cosmoclimatology,' a new theory emerges. Astron Geophys 48(1):1.18–1.24; Svensmark H, Bondo T, Svensmark J (2010) Cosmic ray decreases affect atmospheric aerosols and clouds. Geophys Res Lett 36:L15101

16. Shaviv NJ, Veizer J (2003) Celestial driver of Phanerozoic climate? GSA Today 13(7):4–10

17. Shaviv NJ (2002) Cosmic ray diffusion from the galactic spiral arms, iron meteorites, and a possible climate connection? Phys Rev Lett 89:051102

18. Tinsley BA, Deen GW (1991) Apparent tropospheric response to Mev-Gev particle flux variations: a connection via electrofreezing of supercooled water in high-level clouds. J Geophys Res 96:22283–22296

19. Shaviv NJ (2002) The spiral structure of the Milky Way, cosmic rays, and ice age epochs on Earth. New Astron 8:39–77

20. Hoyle F, Wickramasinghe NC (1978) Comets, ice ages, and ecological catastrophes. Astrophys Space Sci 53:523–526

21. Rahmstorf S, Archer D, Ebel DS, Eugster O, Jouzel J, Maraun D, Neu U, Schmidt GA, Severinghaus J, Weaver AJ, Zachos J (2004) Cosmic rays, carbon dioxide, and climate. Eos 85(4):38; Royer DL, Berner RA, Montañez IP, Tabor NJ, Beerling DJ (2004) CO_2 as a primary driver of Phanerozoic climate. GSA Today 14:4–10; Overholt AC, Melott AL, Pohl M (2009) Testing the link between terrestrial climate change and galactic spiral arm transit. Astron J 705:L101; kfc (2009) Spiral arms did not cause climate change on Earth. Technology Review, Massachusetts Institute of Technology, June 2009

22. Shaviv NJ, Veizer J (2004) Detailed response to "Cosmic rays, carbon dioxide, and climate". Eos 85(4):38

23. Shaviv NJ, Veizer J (2004) Further response to "Cosmic rays, carbon dioxide, and climate. www.phys.huja.ac.il/~shaviv/ClimateDebate/RahmReplyReply.pdf

24. Shaviv N, Veizer J (2004) CO_2 as a primary driver of Phanerozoic climate: Comment. GSA Today; Shaviv NJ (2005) On climate response to changes in the cosmic ray flux and radiative budget. J Geophys Res 110:A08105

25. Bailer-Jones CAL (2009) Evidence for and against astronomical impacts on climate change and mass extinctions: a review. Int J Astrobiol 8:213–239; Usoskin IG, Marsh N, Kovaltsov GA, Mursula K, Gladysheva OG (2004) Latitudinal dependence of low cloud amount on cosmic ray induced ionization. Geophys Res Lett 31(16):L16109–L16110; Usoskin IG, Korte M, Kovaltsov GA (2007) Role of centennial geomagnetic changes in local atmospheric ionization. Geophys Res Lett 35:L05811; Usoskin IG, Desorgher L, Velinov P, Storini M, Flückiger EO, Bütikofer R, Gennady GA (2009) Ionization of the Earth's atmosphere by solar and galactic cosmic rays. Acta Geophys 57(1):88–101; Harrison RG, Stephenson DB (2006) Empiri-

cal evidence for a nonlinear effect of galactic cosmic rays on clouds. Proc R Soc A Math Phys Eng Sci 462(2068):1221–1233

26. Marsh N, Svensmark H, Christiansen F (2005) Influences of solar activity cycle on Earth's climate. February, Publication of the Danish National Space Center

27. Kristjansson JE, Stjern CW, Stordal F, Fjaeraa AM, Myhre G, Jonasson K (2008) Cosmic rays, cloud condensation nuclei and clouds – a reassessment using MODIS data. Atmos Chem Phys 8(24):7373–7387; Kulmala M, Rüpinen I, Neiminen T, Hulkkonen M, Sogacheva L, Manninen HE, Paasonen P, Petäjä T, Dal Maso M, Aalto PP, Viljanen A, Usokin I, Vainio R, Mirme S, Mirme A, Minikin A, Petzold A, Hõrrak U, Plaß-Dülmer C, Birmili W, Kerminen V-M (2010) Atmospheric data over a solar cycle: no correlation between galactic cosmic rays and new particle formation. Atmos Chem Phys 10:1885–1898; Erlykin AD, Sloan T, Wolfendale AW (2010) Cosmic rays and global warming. Europhys News 41(1):27–30. http://dx.doi.org/10.1051/epn/2010104; Calogovic J, Albert C, Arnold F, Beer J, Desorgher L, Flueckiger EO (2010) Sudden cosmic ray decreases: no change of global cloud cover. Geophys Res Lett 37(3):L03802

28. CERN. public.web.cern.ch/public/en/Research/Research-en.html

29. The CLOUD experiment. public.web.cern.ch/public/en/Research/CLOUD-en.html

30. CERN's CLOUD experiment provides unprecedented insight into cloud formation. Press release, 25 Aug 2011. press.web.cern.ch/press/pressreleases/Releases2011/PR15.11E.html

31. CLOUD experiment results. Supporting information: press.web.cern.ch/press/pressreleases/Releases2011/PR15.11E.html (– then:) CLOUD SI press-briefing 29JUL11.pdf

Index

A. Cooke, *Astronomy and the Climate Crisis*,
Astronomers' Universe, DOI 10.1007/978-1-4614-4608-8,
© Springer Science+Business Media New York 2012

Printed by Publishers' Graphics LLC